D0539018

ALSO BY DAVID WEINBERGER

Small Pieces Loosely Joined

The Cluetrain Manifesto
with Christopher Locke, Rick Levine, and Doc Searls

EVERYTHING IS MISCELLANEOUS

EVERYTHING IS MISCELLANEOUS

THE POWER OF THE NEW DIGITAL DISORDER

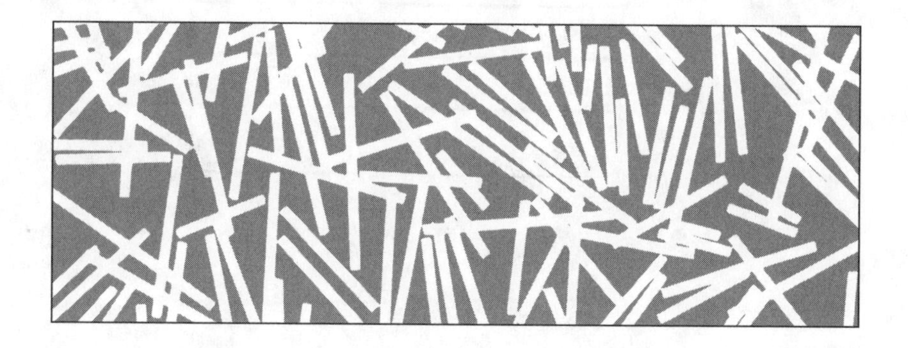

DAVID WEINBERGER

TIMES BOOKS
HENRY HOLT AND COMPANY ■ NEW YORK

Times Books
Henry Holt and Company, LLC
Publishers since 1866
175 Fifth Avenue
New York, New York 10010
www.henryholt.com

Henry Holt® is a registered trademark of
Henry Holt and Company, LLC.

Distributed in Canada by H. B. Fenn and Company Ltd.

Library of Congress Cataloging-in-Publication Data is available.
ISBN-13: 978-0-8050-8043-8
ISBN-10: 0-8050-8043-0

Henry Holt books are available for special promotions and
premiums. For details contact: Director, Special Markets.

First Edition 2007

Designed by Kelly S. Too

Printed in the United States of America
1 3 5 7 9 10 8 6 4 2

To the librarians

CONTENTS

INFORMATION IN SPACE

"Absolutely not."

I've apparently begun by asking Bob Medill the wrong question: "Don't you put the most popular items in the back?" He could have taken it as an insult, for it's a customer-hostile technique many retailers use to force shoppers to walk past items they hope they'll buy on impulse. But the soft-spoken Medill is confident in his beliefs. Besides, he's been asked that before. It's a rookie question.

"No," he says, looking out over the Staples office supply store he manages. "In front are the destination categories because that's what our customers told us they want." His arm sweeps from left to right, gesturing to the arc of major sections of the store: "Paper, digital imaging, ink and toner, business machines, and the copy center."

It's two o'clock in the afternoon, but we have the place to ourselves. Even if a customer wanted to buy something, no one is at the cash register. If you need help with your purchase, no "associates"—Staplesese for "sales assistants"—are available. Medill is unconcerned. That's the way it's supposed to be. We're in the Prototype Lab, a full-sized store mock-up at the company's headquarters in an office park in Framingham, Massachusetts.

The site has nothing of a Hollywood set about it. It's all real and fully stocked, from the twenty-four-pound paper marked on sale to the blister-packed pens hanging neatly side by side. Eight people work there full-time, which is less than a real store's typical complement of twenty-nine but still no small expense. Yet it's worth it because, despite the aisles of pens and the pallets of paper positioned by forklifts, the Prototype Lab is actually about information. Every day Bob Medill and his staff work on strategies to overcome the limitations of atoms and space so customers can navigate a Staples store as if it were pure information.

That's not the way Medill would put it. From his point of view, the Prototype Lab is a testing ground for making shopping at Staples easier for customers. That by itself puts him in the vanguard of merchandisers. More typical merchandisers use physical space against customers so that customers will spend more money than they intend. It's a science retailers know well. Supermarkets stock popular items, such as milk and bananas, in the back of the store to take advantage of the way physical space works: To get from area A to aisle C, we have to go past shelf B, which just happens to have a sign announcing a special on something we didn't come in for. Likewise, you'll find doggie treats below eye level because it's something kids are more likely than their parents to put in the cart. When Medill talks about making it easier for Staples' customers to get out of the store fast, he's a bona fide revolutionary.

"Customers fall into two buckets," says Liz McGowan, Staples' director of visual merchandising. "People who feel that asking for help is a personal failure and those who don't." Despite what comedians tell us, the dividing line is not based on gender. "My mother is in the first bucket," she says. McGowan is data-driven, so she knows the precise volume of the buckets. "Thirty-two percent ask associates. Twenty-four percent use signage. Forty percent already know where things are." It's the 60 percent who need help that determine the informational layout of the store. In the Prototype Lab, that's known as "way-finding," and it's where how people think meets the way their bodies deal with space.

"We learn by watching our customers' eyeballs," Medill says. Customers enter the store and move nine to twelve feet in, and then they—we—"stand and scan." That's why, unlike most stores, Staples doesn't put much signage in the entranceway. Instead, it places signs over the most popular destinations, and signs for subcategories under those signs, like a map of continents divided into countries and then into states. Gesturing at the cleanliness of the design, Medill says, "Originally we had 'focals,' "—signs that call out special offers—"but they blocked eyeballs." In the retail world, the point of "focals" is to interrupt the logical order of the store, bringing some exceptional, can't-be-missed offer to your attention. But focals are also concrete objects, so they not only grab your attention, they also physically obscure information about the store, like a map that puts a big "McDonald's here!" label that obscures most of downtown Poughkeepsie. That's just the way eyeballs work. Because a sign is not information if it can't be physically seen, the average height of human eyeballs also determines the height of the shelves. "By having a store that's mostly low, it's easily scannable," says Medill.

Eyeballs also determine how much information goes on the product description placards that line the shelves, prefacing the products themselves. "With twenty-twenty vision, you have to be able to read it one and a half feet away," explains McGowan. "Three bullets is pretty good," adds Medill. "Five is too many." If human visual acuity were better, there would be more information on the signs, and if we mixed our genes with giraffe DNA, the shelves would be twenty feet tall. And if the shelves were twenty feet tall, a typical Staples might be able to stock 15,000 items instead of 7,200. But why dream? Physical stores are laid out for a species that rarely has eyeballs more than six feet off the ground.

In a physical store, ease of access to information can be measured with a pedometer, and each step is precious. "People come in with lots of ways of identifying printer ink," Medill says. "An old cartridge, an ID number, a printer number, a label from the box." Staples created a catalog of all available printer inks, and gave it its own attractive kiosk. Yet only 7 percent of customers used it. "It was too far away from the inks," Medill explains. "Now we've broken the catalog into pieces and embedded each piece with the relevant merchandise." If you have an Epson printer, you'll find the catalog of Epson inks next to the Epson segment of the ink shelves. "Once we integrated the catalog, twenty percent used it," reports McGowan, the keeper of the numbers.

The purely informational layout of the Prototype Lab is warped by the brute fact that in the physical world, two objects cannot occupy the same space at the same time. As we arrange items in space, we're also determining the time it will take to reach them. Eliminate this basic fact of the physical world and there'd be no need for the Prototype Lab.

Of course, we could try stocking the same item in many places throughout the store. But most stores, including Staples, don't like to do that. I ask Mike Moran, the person in charge of figuring out the spatial relationships, for an example of an item that's stocked in more than one place. "Cables," he responds immediately. "What do I use them for? For printing," he says, assuming the customer's point of view. So cables are in with printers. But they're also in a separate cable section. Yet the same argument could be made for stocking blank CDs and DVDs in lots of places. Instead they're confined to a display toward the start of the arc of destination areas Medill had gestured to. Why aren't they also stocked next to the devices that record onto them? Why not also next to paper, since both are ways of recording information? Why not also with software, since they're both CDs? For that matter, why not put pens with paper, with notebooks, with the yellow stickies, and with the blank labels? "Operational simplicity," says Moran. If CDs were put everywhere a customer might want to find them, it would be impossible to make sure that each pile was kept stocked. Besides, it would eat up shelf space, a commodity so limited that in groceries and bookstores, vendors pay for the privilege of having their goods placed well. Destination areas are the only places where there's double stock because, Moran says, "If I leave the store with a printer but not cables, paper, and ink, the product isn't usable, and I come back annoyed."

Having to come back: the victory of space and time over the human ability to remember what goes with what. Many of us find it unreasonably irritating to have to make a second trip to pick up what we forgot the first time—what we forgot because the store-as-information failed to help us remember. Information is easy. Space, time, and atoms are hard.

Medill's crew doesn't think of it this way, but they're in a battle. Their constant enemy is the physical, three-dimensional world itself. Software programmers would say that the people at the Prototype Lab are "hacking the physical"—finding clever ways around the limitations built into the system. The limitations are so much a part of our everyday world that we don't even recognize them as such. For example:

In physical space, some things are nearer than others. That's why Liz McGowan worries about way-finding: She wants us to be able to get everything on our shopping list with the minimum number of steps.

Physical objects can be in only one spot at any one time, so McGowan and Moran have to figure out which one place—or two at the maximum—to put items, even though it'd be easier for customers if anything they wanted was always within arm's reach.

Physical space is shared, so there can be only one layout, even though we all have different needs. If you're in a wheelchair, McGowan's careful organization of signs at average height isn't going to work very well for you. Or if you go to Staples primarily for school supplies, you'll probably find the store's choice of what counts as a destination area irrelevant, since it doesn't include crayons and three-hole Harry Potter notebooks.

Human physical abilities are limited, so the amount of information provided to us is constrained by our ability to see; you wouldn't want the informational signs to be so detailed that they obscured the products themselves.

The organization of the store needs to be orderly and neat. If things are out of place, they can't be found; the physical mapping of the store needs to reflect the organization of the information, and that organization needs to be as simple as possible. A messy store is a disordered store is a failed store.

These limitations mean that no matter how well Medill and his group do their jobs, most of what's in Staples is just in our way. If I come in with a shopping list of fifteen items, the other 7,185 items Staples stocks not only are irrelevant, they hide what I'm looking for. If, magically, those fifteen items were all that was in the store when I got there—and that was true each time I came, no matter what was on my list—I could go to a single shelf at the front of the store, sweep them into my basket, and be done with it. There wouldn't be any need for McGowan's way-finding studies or for Moran's careful consideration of what should be placed next to what.

But we all know how reality works, so why worry about what might be possible in some sci-fi alternative universe?

Because the alternative universe exists. Every day, more of our life is lived there. It's called the digital world.

Instead of atoms that take up room, it's made of bits.

Instead of making us walk long aisles, in the digital world everything is only a few clicks away.

Instead of having to be the same way for all people, it can instantly rearrange itself for each person and each person's current task.

Instead of being limited by space and operational simplicity in the number of items it can stock, the digital world can include every item and variation the buyers at Staples could possibly want.

Instead of items being placed in one area of the store, or occasionally in two, they can be classified in every different category in which users might conceivably expect to find them.

Instead of living in the neat, ordered shelves we find in the Prototype Labs, items can be jumbled digitally and sorted out only when and how a user wants to look for them.

Those differences are significant. But they're just the starting point. For something much larger is at stake than how we lay out our stores. The physical limitations that silently guide the organization of an office supply store also guide how we organize our businesses, our government, our schools. They have guided—and limited—how we organize knowledge itself. From management structures to encyclopedias, to the courses of study we put our children through, to the way we decide what's worth believing, we have organized our ideas with principles designed for use in a world limited by the laws of physics.

Suppose that now, for the first time in history, we are able to arrange our concepts without the silent limitations of the physical. How might our ideas, organizations, and knowledge itself change?

That journey will take us from Aristotle to the quiet psychology professor in Berkeley, California, who proved him wrong. From scientists trying to number living things to the businesses that are deciding that if they make their information messier, it'll be easier to find. From the eighteenth-century encyclopedists who were accused of violating God's order because they arranged topics alphabetically to the world's first encyclopedia without editors, page limits, or order.

And here's a hint about what we will find. As we invent new principles of organization that make sense in a world of knowledge freed from physical constraints, information doesn't just want to be free. It wants to be *miscellaneous*.

THE NEW ORDER OF ORDER

Before the Web, the word *browsing* was usually a polite way of telling a salesperson to buzz off. "May I help you?" a salesperson asks. "I'm just browsing," you reply with a little smile. With that word, a customer declares a lack of commitment. With that smile, she asserts that she's within her rights: "Just try and stop me, salesboy!"

Browsing is more than window-shopping, fantasizing about what it would be like to own something or resenting those who do. You browse when you intentionally ignore the organizational structure the store has carefully imposed on its stock. You have a hankering to read something about the Civil War, but the bookstore has Civil War books strewn about the fiction, nonfiction, biography, and travel sections, all neatly arranged in individual aisles and on individual shelves. Or you're in the mood for some light reading—nothing more specific than that—and appealing books pop up on just about every shelf. The store helps you violate its order by providing tables at the front with staff picks, new books, and bargains, but it can't anticipate all the ways every customer who walks in the door is going to want to browse. So it has to depend on its sales staff to help its customers find the exact book they want when they say, "I need something as a sweet sixteen present." Sometimes they get it right; sometimes—say, if the available staff person thinks all the kids still have a favorite Beatle—they lose the sale as the customer decides it would be safer just to give her niece a check.

The normal organization of a store works well enough if you come in knowing what you want: Go to the fiction shelf, find the "A" section conveniently located at the beginning of the alphabetized authors, and locate that copy of *Pride and Prejudice* for your niece. But *discovering* what you want is at least as important as *finding* what you know you want. Our bookstores look like they prefer seekers over browsers because the usual layout works well for people trying to find what they came in for, whereas there are almost as many ways to organize for browsers as there are browsers. An order that works for one interest may not for others—clustering all the books about the Civil War would help the Civil War buffs but would pull *Gone with the Wind* off the shelf where the historical fiction buffs browse. On the other hand, dumping the shelves into a bookstore-sized pile of books would turn browsing into pawing through printed rubble.

If only there were a way to arrange the stuff in stores so that every possible interest could be captured. When we know what we want, we'd find it immediately. When we want to browse, the store would rearrange itself based on our needs and interests, even when we aren't quite sure what those are.

At Apple Computer's iTunes music store, it's already happened. For decades we've been buying albums. We thought it was for artistic reasons, but it was really because the economics of the physical world required it: Bundling songs into long-playing albums lowered the production, marketing, and distribution costs because there were fewer records to make, ship, shelve, categorize, alphabetize, and inventory. As soon as music went digital, we learned that the natural unit of music is the track. Thus was iTunes born, a miscellaneous pile of 3.5 million songs from over a thousand record labels. Anyone can offer music there without first having to get the permission of a record executive. Apple lets customers organize the pile any way they want and markets *through* their customers' choice of tracks and playlists rather than *to* the mass market. By making music miscellaneous, Apple has captured more than 70 percent of the market.

And the iTunes store isn't even all that miscellaneous. It's a spreadsheet that can be sorted by the criteria iTunes provides: the track's name, length, artist, album, genre, and price. If you want to browse, first you pick the genre, artist, and album, in that order. If you want to browse by the artist and then by genre, you can't. If you want to browse by mood, language, or date, you can't. Even though iTunes is all digital, there are still more ways you can organize your physical collection of CDs. The problem isn't that iTunes has chosen an inappropriate set of criteria for sorting, although that certainly could be argued. The real problem is that iTunes accepts the premise we've had to operate under in the physical world: that there *is* a set of appropriate criteria.

While iTunes is parsimonious in its built-in ways of sorting, it generously enables customers to create their own playlists, pulling together songs from across the entire ocean of tunes. By allowing customers to then publish their playlists—and rate and comment on other people's—iTunes provides as many ways to navigate its inventory as there are customer moods and interests. That lets in an important breath of the miscellaneous and shows that iTunes has learned the right lesson: To get as good at browsing as we are at finding—and to take full advantage of the digital opportunity—we have to get rid of the idea that there's a best way of organizing the world.

EVERYTHING HAS ITS PLACES

It won't be easy. The world started out miscellaneous but it didn't stay that way, because we work so damn hard at straightening it up. Take eating, the most basic bodily activity we do on purpose. Preparing to serve a meal is a complex dance of order. We have separate storage areas for each of the implements of eating: silverware, plates, glasses, napkins. Each of these areas has its own principle of organization: plates stacked by type, clustered by pattern and how "good" they are, juice glasses separated from wineglasses and from tumblers, silverware tucked cozily into special compartments. We then break these items apart to form patterns (salad fork to the left of dinner fork) determined by tradition, one set per participant in the meal. When we clear the table, we recluster the items by type because we generally don't wash the implements in the same pattern we used when we were eating with them: We do all the plates at once, put the silverware into a pot to soak, and line up the glasses next to the sink. If we stack the dishwasher, we cluster the items yet another way. When it's all clean and dry, we again store everything in its initial order, completing our nightly choreography.

We juggle multiple principles of organization without even thinking about it. You know what goes in your spice rack and what doesn't, even though the principle of order is hard to find: What makes dried leaves (oregano), dried seeds (nutmeg), and dried bark (cinnamon) all spices? All add a little more flavor to a dish? But so do chocolate sprinkles, and they don't count as a spice, despite coming in a shaker the size and shape of an oregano bottle. And even if you count salt and black pepper as spices, you probably don't keep them in the spice rack, because you use them too often. Without pausing for thought, you have coordinated four intersecting sets of criteria: how big the bottles are, what the contents are used for, which part of a meal they're applied to, and how frequently you need them.

The same is true for every room, every closet, and every tabletop in our houses. Even the oddest, most random items have their place. The gin-flavored floss someone gave you as a joke goes on the top shelf of the bathroom cabinet, while the Miss Piggy night-light goes in with the rarely used electrical equipment in the box in the basement. If you're genuinely stumped, you'll probably throw it in a box of things to give away.

The two processes by which new things are introduced into our homes are typical of how we handle information: We *go through* new arrivals and then we *put them away*. We go through the mail and file it in the special places we have for bills (the desk), cards from relatives (the refrigerator door), and junk mail (the trash). We go through bags of groceries and put the food away within minutes of bringing it into our house. We address these elements of disorder—unsorted mail in the mailbox, groceries sorted by relative weight into bags by a clerk in the store—with remarkable alacrity.

There isn't a part of our homes that is truly unordered, except perhaps under our beds, and for many of us even that is the site of the spontaneous ordering of dust into bunnies.

We invest so much time in making sure our world isn't miscellaneous in part because disorder is inefficient—"Anybody see the gas bill?"—but also because it feels bad. Knowing where things are and where things go is essential to feeling at home. If cleanliness is next to godliness, then slovenliness is next to *The Odd Couple*'s Oscar Madison. And who wants to be next to Oscar Madison?

We've been raised as experts at keeping our physical environment well ordered, but our homespun ways of maintaining order are going to break—they're already breaking—in the digital world. The most visible breach so far: the folder on the family computer that stores the digital photos.

If you're managing to keep your digital photos well organized, congratulations. But you're probably going to lose the battle sooner rather than later. It's a simple matter of numbers. A typical album you buy at your local camera store holds between fifty and two hundred paper-based photos. You likely have a thousand photos in all your albums put together. A thousand photos, each with its own story. "Oh, remember how Mimi was always wearing that silly cowboy hat? And there's Aunt Sally on the beach. She was so sunburned we had to take her to the hospital, and that funny doctor said we should just baste her in barbecue sauce."

Now check your computer. If you have a digital camera, you may well have saved over a thousand photos in just the past few years. It's only going to get worse. Digital cameras started outselling film cameras in the United States in 2003 and worldwide in 2004. And, in 2004, 150 million cell phones with cameras were sold, almost four times the number of digital cameras. Because digital photos are virtually free, we're tempted to take more and more pictures, sometimes just in the hope that one will come out well. We're also keeping more of the photos, and not always because we want them: Since our cameras apply names like "DSC00165.jpg" to our photos, it's easier to keep bad photos than to throw them out. To keep them, we just press a button to move them from our camera. To get rid of them, we have to look at each one, compare it with the others in the series, select the bad ones, press the Delete button, and then confirm our choice.

As a result, we are loading onto our computers thousands of photos with automatically generated names that mean nothing to us. When you have ten, twenty, or thirty thousand photos on your computer, storing a photo of Aunt Sally labeled "DSC00165.jpg" is functionally the same as throwing it out, because you'll never find it again.

We're simply not going to be able to keep up. Even obsessive-compulsives have only twenty-four hours in a day. Perhaps technology will get better at automatically figuring out what and who is shown in a photo. Or perhaps labeling photos will become a social process, with others pitching in to help us organize them. The user-based organizing of photos is already happening on a massive scale at Internet sites like Flickr.com, where people can post their photos and easily label them, allowing others to search for them. Moreover, anyone can apply descriptive labels to photos and create virtual albums made up of photos taken by themselves and strangers. What's clear is that however we solve the photo crisis, it will be by adding more information to images, because *the solution to the overabundance of information is more information.*

We add information in the real world by putting a descriptive sign on the shelf beneath a product, sticking a label on a folder, or using a highlighting pen to mark the passages that we think will be on the test. The real world, though, limits the amount of additional data we can supply: Staples has to keep the product information labels on the shelves small enough so they won't obscure the product; a manila folder's label can't have more than a few dozen characters on it without becoming illegible; and if previous students have already highlighted every other sentence in your textbook, the marks you make won't add much information at all. In the digital world, these restrictions don't hold. The product listing on the Staples Web site can link to entire volumes of information, our computers can store more information about a desktop folder than is actually in the folder, and if the digital textbook has had every word highlighted by previous readers, a computer could show us which sections have been highlighted by the majority of A students.

Such features are not just cool tricks. They change the basic rules of order. When we come across the paper photo from 2005 of Aunt Sally on a beach in Mexico at sunset celebrating cousin Jamie's birthday, with the twins in the background playing badminton, we have to decide which one spot in one album we're going to stick it into. If it were a digital album, we wouldn't have to make that choice. We could label it in as many ways as we could think of: Aunt Sally, Mexico, 2005, beach, birthday, twins, badminton, sunset, trips, foreign countries, fun times, relatives, places we want to go back to, days we got sunburned. That way we could have the computer assemble albums based on our interests at the moment: all the photos of the cousins, all the trip photos for the past five years, all the photos of Aunt Sally having fun. The digital world thereby allows us to transcend the most fundamental rule of ordering the real world: Instead of everything having its place, it's better if things can get assigned multiple places simultaneously.

YOURS, MINE, OURS

These types of changes create effects that are rippling through our social world. Recently, my sister-in-law organized some of our parents-in-law's physical photos into a traditional album as an anniversary gift from all of us. She balanced the pictures so that her kids didn't dominate, remembered to include photos of dear friends even if the only available snaps were rather unflattering, and carefully placed them in an album in chronological order. Had she gone off and done this without asking, it would have been highly presumptive because building an album is often a ritual a family shares. We hand the pictures around, clucking at silly expressions and worse haircuts, laughing at the escapades the photos capture. Together we construct our past for the future, making the decisions about which photographs to put next to which, "chunking" the smoothness of experience into lumps of meaningful memories.

We do something quite different with our digital photos. A digital album is the same as an iPod music playlist: a way to remember a particular arrangement of photos. A single photo can go on dozens of playlists at virtually no cost. So if my sister-in-law takes five hundred photos during her trip to Paris, she can make one digital album that focuses on her kids' reactions, another of interesting faces of Parisians, and another of every plate of food she ate. If she chooses to share her digital photos with her extended family, perhaps one of us will cluster the photos of public art or of her kids making faces. There's no limit to how many albums we can assemble. So, we're no longer forced to carefully construct a single shared path through memory. Rather, the more ways our digital photos can be sorted, ordered, clustered, and made sense of—the more miscellaneous they are—the better. We lose the requirement that a family get on the same page (literally) about its memories. And if albums are the archetypes of memory, memory becomes less what we *have* assembled and locked away and more what we *can* assemble and share.

The changes we're facing are not just personal. We have major institutions dedicated to keeping the world from slipping into the miscellaneous. The Library of Congress owns 130 million items, including 29 million books on 530 miles of shelves. Every *day,* more books come into the library than the 6,487 volumes Thomas Jefferson donated in 1815 to kick-start the collection after the British burned the place down. The incoming books are quickly sorted into cardboard boxes by topic. Then the boxes are delivered to three to four hundred catalogers, who between them represent eighty different subject specializations. They examine each book to see which of the library's 285,000 subject headings is most appropriate. Books can be assigned up to ten different subject headings. Keeping America's books nonmiscellaneous is a big job.

So is building and maintaining the subject headings themselves. "We create eight thousand new subject headings per year, with about as many adjustments to existing subject headings," says Barbara Tillett, chief of cataloging policy. The structure of the library's catalog is open-ended enough to allow a subject heading to be created to accommodate books on new topics, rather than insisting on cramming books into the existing headings. The library is also willing to modify the system's larger-scale arrangement, as when the new category of Environmental Sciences was created because, says Tillett, catalogers were bouncing books back and forth: "This one is yours." "No, it's yours."

Catalogers, using their years of experience, are free to propose new subject headings for an up-or-down vote by a committee composed of senior librarians. At one typical weekly meeting, just one out of eighty proposed classifications was turned down—a subclassification of the Photographic Portraits subheading to cover Sri Lankans. The committee decided that that heading's subclassifications were intended to classify people by type—children, physicians, and so on—rather than by country of origin. Such issues may not be glamorous, but they are the type faced daily by those patrolling the conceptual borders of book categorization.

It takes hundreds of professionals with centuries of cumulative expertise to keep the Library of Congress well ordered. But even though the Library of Congress has itself become a standard unit of measurement for large objects—for example, NASA says it maintains information about the environment that would "fill the Library of Congress 300 times"—it's only dealing with seven thousand new books a day. The *Washington Post* estimates that *seven million* pages are added to the Web every day. Search on Google for "American history," which is just one Library of Congress subheading, and you'll get 750 million Web pages—about twenty-six times the number of books in the Library's entire book collection. The Library of Congress's carefully engineered, highly evolved processes for ordering information simply won't work in the new world of digital information. Not only is there too much information moving too rapidly, there are no centralized classification experts in charge of the new digital world we're rapidly creating for ourselves, starting with the World Wide Web but including every connected corporate library, data repository, and media player.

If the Library of Congress's well-proven approach won't work as we digitize our information, ideas, and knowledge, what will?

THE THREE ORDERS OF ORDER

Bill Gates bought the Bettmann Archive, the most prestigious collection of historic photos in the United States, so he could bury it. In 2001 he hired nineteen trucks to move it from the melting summers of Manhattan to a cool limestone cave 220 feet underground in the middle of Pennsylvania. There, dehumidifiers the size of closets hold down the moisture level, and security guards patrol brightly lit streets carved out of stone. The site, run by the records management company Iron Mountain, is modern in every way, but if you walk far enough you'll come to a dead end where a hole in the wall reveals an underground lake illuminated only by the light from the opening through which you're looking.

The photographs in the Bettmann Archive are stored in a long narrow cavern whose arched walls of rough rock have been painted white but otherwise left unfinished. In rows of filing cabinets that stretch to the vanishing point are 11 million priceless photographs and negatives. They are arranged by the originating collections the Bettmann purchased over time. Within the collections the photos and negatives are generally ordered chronologically. The room is being slowly lowered to −4° Fahrenheit because Henry Wilhelm, a leading authority on film preservation, believes that at that temperature it will take five hundred years for the collection to deteriorate as much as it did in a single year when it was kept in Manhattan. Wilhelm was inspired to make film preservation his life's work when, as a member of the Peace Corps deep in the Bolivian rain forest, he saw treasured family photographs in thatched houses. "They were deteriorating badly, and there was nothing I could do about it," he says, still sounding frustrated decades later. The Bettmann facility he designed is the polar opposite of a Bolivian rain forest.

As you stand in the long cavern, you are in the midst of a huge first-order organization. In the first order of order, we organize things themselves—we put silverware into drawers, books on shelves, photos into albums. But when you go through the air lock that Wilhelm designed to connect the back chamber and the front one, you confront a prototypical example of the second order of order: a card catalog containing information about each of the eleven million objects in the back cavern. The catalog separates information about the first-order objects from the objects themselves, listing entries alphabetically by subject so that you can find, say, all the photos of soldiers across all of the archive's collections. A code on this second-order object, the catalog card, points to the physical place where the first-order photo is stored in the back room. But quite a few of the Bettmann's photographs are not listed in the card catalog. Some of the older collections arrived with catalogs entered in hand-written ledger books, one line per photo, listed in the order in which the photo was received. Finding a photo in one of those collections requires looking through the ledgers' yellowing pages line by line, hoping to come across a description of the image you're seeking. The ledgers are also a form of second-order classifications, just a much less efficient method than the card catalog.

The Bettmann's second-order organization works, but it's expensive to maintain, and retrieval times are sometimes measured in days. And there are limits inherent in the second order. Not all the information about the objects is recorded; a photograph of a Massachusetts soldier in the Civil War eating in a field, his rifle by his side, might be listed under "Civil War" and "soldier," but probably not also under "Massachusetts," "rifle," "weapons," "uniforms," "dinner," and "outdoors." That means if you were to ask the Bettmann's curators if they had a photo of a Civil War soldier eating outdoors, they would have to send someone into the stacks and stacks of filing cabinets to do a search through the photos themselves. Even if all that data were recorded, it would swell the size of the card catalog to the point of unusability: Searching through eleven million cards at one per second would take over four months of round-the-clock riffling.

All that work—a long line of trucks to move the archive, a hole dug deep into the earth, an ambient temperature growing so cold that you have to don arctic gear to enter the vault—and we still get so little use of those valuable—and expensive—assets. Indeed, a first- and second-order archive the size of the Bettmann literally cannot know everything it has.

The problems with the first two orders of order go back to the fact that they arrange *atoms*. There are laws about how atoms work. Things made of atoms tend to be unstable over time—paper yellows and disintegrates, negatives turn to soup—so we have to take measures to sway nature from its course. Atoms take up room, so collections of photos can get so large that we have to build card catalogs to remind us of where each photo is. And things made of atoms can be in only one spot at a time, so we have to decide whether a photo of a soldier eating should go into the Civil War folder or the Outdoor Meals folder.

But now we have bits. Content is digitized into bits, and the information about that content consists of bits as well. This is the third order of order and it's hitting us—to use a completely inappropriate metaphor—like a ton of bricks. The third order removes the limitations we've assumed were inevitable in how we organize information.

For example, the digital order ignores the paper order's requirement that labels be smaller than the things they're labeling. An online "catalog card" listing a book for sale can contain—or link to—as much information as the seller wants, including user ratings, the author's biography, and the full text of reviews. You can even let users search for a book by typing in any phrase they remember from it—"What's the title of that detective novel where someone was described as having a face like a fist?"—which is like using the entire contents of the book as a label. That makes no sense when all that information has to be stored as atoms in the physical world but perfect sense when it's available as bits and bytes in the digital realm.

You can see the third order in action by flying across the country from the Bettmann Archive to Seattle, where Corbis, the Bettmann's parent company, has its headquarters. Corbis has charmingly renovated an old bank, knocking down walls to let in light and air, and even retaining the old circular vault door, symbolically open and inviting. Corbis holds over four million digital images, a collection smaller than the Bettmann's but subject to the same issues of organization and control. Because the images as well as the information about the images are all fully electronic, Corbis organizes its photos without regard to the physical constraints that limit the curators back at the Bettmann. At Corbis, you can find a digital image of a Civil War soldier eating dinner by typing "soldier," "Civil War," and "meal" into the search engine, or by browsing a list of categories and subcategories. You can find what you need in seconds. If you don't, you can be pretty sure it just isn't in the Corbis collection.

Of course, it took Corbis many hours of preparation to reduce these search times to seconds. A team of nine full-time catalogers categorize each image Corbis acquires, anticipating how users are going to search. When a new image comes into the collection, one of the catalogers uses special software to browse the 61,000 "preferred terms" in the Corbis thesaurus for those that best describe the content of the image, typically attaching 10 to 30 terms to each one. The system incorporates about 33,000 synonyms (searches on "beach" turn up images labeled as "seaside"), as well as more than 500,000 permutations of names of people, movies, artworks, places, and more. That broadens the side of the barn so wide that if you misspell Katharine Hepburn's name as Katherine or Catherine, you'll still find all the images of the high-cheekboned screen legend in the Corbis collection. And if you're looking for Muammar Gadhafi, at least seventeen different ways of spelling his name will get you what you want.

The differences between the Bettmann's second-order organization and Corbis's third-order method affect every aspect of their businesses.

The Bettmann is an attic that's never been fully explored. It doesn't know all the photos it owns; a ledger entry may be buried among thousands of others, and it may not describe the photo in a way that enables people who want it to find it. At Corbis, every image has been carefully cataloged and can be found by using the company's search engine.

The Bettmann has to be parsimonious with its information: Create too many catalog cards for each photo and your card catalog bloats to unriffability. Like Staples, it bumps against the limits of the physical world. Corbis's approach to information is sprawling and extravagant: The immateriality of bits encourages Corbis to put its images in every place where people might look for them.

Because the Bettmann collection's second-order information—known as *metadata* because it's information about information—is incomplete and spread across catalogs and ledgers, it's accessible to only a handful of trained experts. Corbis, because its collection is third-order and thus fully digital and cataloged, is designed to be searched by any customer.

Finding photos among the Bettmann's assets can be a slow, manual, expensive process. Corbis, on the other hand, not only can derive benefit from every image but knows so much about them that it can offload much of the job of searching for photos to its users.

Corbis's digital images do not deteriorate with age and require far less physical maintenance. Overall, Corbis spends less per image, can make more per image, and is able to turn more of its images into productive assets.

The differences in the order of order even drive differences in the lighting. The Bettmann's archives are brightly lit because their images are made of atoms that are visible only when light reflects off them. Corbis's catalogers work in semidarkness because digital images on monitors are their own source of light.

But here's the kicker: Like the iTunes Store, Corbis isn't even a particularly good example of third-order organization. It's doing what's right for its business at the moment, but it's still doing the basic second-order task of having professionals stick things into folders. Granted, the things and the folders are electronic, so Corbis can get more value from its assets at a lower cost. But there are other organizations that are able to move further down the third-order path. Corbis gives us only a taste of the revolution that's under way. Just take a look at Flickr to see one way this is unrolling. With over 225 million photos already uploaded by users and almost a million added every day, Flickr's collection dwarfs that of Corbis and Bettmann. Flickr has no professional catalogers. It relies solely on the labels users make up for themselves, without control or guidance. Yet it is remarkably easy to find photos at Flickr on almost any topic and to pull together collections of photos on themes that mix and match those topics at will. Want to find photos of dogs wearing red clown noses? A search at Flickr finds nineteen of them. Researching car-crash art? Flickr finds thirty-two photos that may help your studies.

The digital revolution in organization sweeps beyond how we find odd photos and beyond how we organize our businesses' information assets. In fact, the third-order practices that make a company's existing assets more profitable, increase customer loyalty, and seriously reduce costs are the Trojan horse of the information age. As we all get used to them, third-order practices undermine some of our most deeply ingrained ways of thinking about the world and our knowledge of it.

For example, medical information that used to come only through the careful filters of medical experts and medical publications is now available to everyone prior to the basic housekeeping processes of being *gone through* and *put away*. The miscellanizing of this information not only breaks it out of its traditional organizational categories but also removes the implicit authority granted by being published in the paper world. Second-order organization, it turns out, is often as much about authority as about making things easier to find.

We have entire industries and institutions built on the fact that the paper order severely limits how things can be organized. Museums, educational curricula, newspapers, the travel industry, and television schedules are all based on the assumption that in the second-order world, we need experts to go through information, ideas, and knowledge and put them neatly away.

But now we—the customers, the employees, anyone—can route around the second order. We can confront the miscellaneous directly in all its unfulfilled glory. We can do it ourselves and, more significantly, we can do it together, figuring out the arrangements that make sense for us now and the new arrangements that make sense a minute later. Not only can we find what we need faster, but traditional authorities cannot maintain themselves by insisting that we have to go to them. The miscellaneous order is not transforming only business. It is changing how we think the world itself is organized and—perhaps more important—who we think has the authority to tell us so.

ALPHABETIZATION AND ITS DISCONTENTS

If you've ever tried to type an Å on an English keyboard, you appreciate the impulse to create a single universal alphabet. Especially as businesses—and their computer systems—go global, having inconsistent alphabets seems to make as much sense as having inconsistent calendars.

In fact, a universal alphabet is such a good idea that we have it about once a generation. It became particularly popular once telegraphs connected the four corners of the earth, removing distance as a barrier and leaving only language itself as an obstacle. In 1879, a German priest, Johann Martin Schleyer, created a universal alphabet and a universal language—Volapük—to go with it. After some initial success, the effort petered out in petty internal political struggles. Of course, it didn't help that the Volapük translation of "Our Father, who art in heaven" began "O Fat obas," or that the language had the word *pük* in its name. Others picked up the struggle after Volapük failed, introducing bills in Congress to create a universal alphabet in 1888, 1901, and 1911.

Then in 1918, Charles Luthy, an obscure eccentric, announced that he'd diagnosed the problem: The previous attempts invented alphabets willy-nilly and asked everyone to agree on them. That, wrote Luthy, "is a great mistake." Luthy gives away the results of his twenty years of labor right in the title of his book:

THE

UNIVERSAL ALPHABET

THE ALPHABET WHICH THE FACTORS THAT
HAVE EVOLVED IN THE PROCESS OF NATURE
LOGICALLY CONSPIRE TO PRODUCE

This Alphabet Is Based Upon The Correct Analysis Of
The Human Speech Sounds, The Correct Analysis Of
The Roman Script Letters (The Writing That
Must Obtain For All Time,) The Unmistakable
Trend Of The English Language Becom-
Ing The Universal Language And Upon
The Handing Down To Posterity
The Vast English Literature
In Its Most Readable Form

It Contains

An Appropriate Letter For Each Of The Forty-Three
Different Speech Sounds In The Human Voice So That
The Alphabet Is Adapted For The Use Of All Nations,
And Will Perfect The Spelling In All Languages

For Luthy the Universal Alphabet—a series of weird additions to the English character set—wasn't just a convenient invention. Rather, "the Universal Alphabet . . . exists in the very nature of things." He based it on the Roman character set, which he believed was founded on principles "as enduring as are the principles of Euclid." In fact, his seven years investigating handwriting led him to see that "Roman script is *natural,* that it is the fittest and the only correct writing, and that all other systems of script must succumb to it." All hail the Caesar of the alphabet!

Luthy didn't think that the *order* of the letters was natural. In his masterwork he perfunctorily announces that the basic English ordering will be maintained, with his proposed new letters inserted after the letters of which they are variations, with detailed, unexplained exceptions ("the *ōā* (*û*) . . . is located as if it were represented by an *o* letter but is represented by an *u* letter"). Thus, even for Luthy, the most extreme proponent of bringing rationality to the alphabet, alphabetical order remains the very model of an arbitrary order. It tells us exactly nothing about the real relationships among the parts. Indeed, its arbitrariness is its virtue: On a field trip, no one gets upset when students are told that A through M go on Bus No. 1 and the N–Zs go on Bus No. 2, but it would be front-page news if students were divided by race, prettiness, or their parents' incomes.

Precisely because alphabetical order is unnatural and arbitrary, it took a long time to be accepted. The foremost historian of alphabetization, Lloyd W. Daly, has found an inscription on the Greek island of Kos, possibly from the third century B.C.E., that breaks into three alphabetized lists 150 names of the participants in the cult of Apollo and Heracles. But alphabetization did not stick. The Romans took their alphabet seriously—they inserted the new letter *G* in the place of *Z*, and then later decided they wanted *Z* back, forcing it to the end of the line—but did not follow the Greeks in using alphabetical order to sort items.

Post-Roman culture in the West kept reinventing alphabetization, and rediscovering that long lists need to be sorted by more than just the first letter. Daly points to a work by Galen in the first century A.D. as the first example of alphabetization that did more than indiscriminately lump together words that start with the same letter. The next reference Daly finds is in the ninth century: Photius of Constantinople criticizing the work of a fifth-century grammarian for alphabetizing only according to the initial letters. A book in 1053 carefully explains how alphabetization works, indicating along the way that it was not in common use. A treatise on Latin written in 1286, the *Catholicon,* by Giovanni di Genoa, spends about four hundred words explaining the process quite precisely, with many examples, at the end of which the author devoutly says, "Now I have devised this order at the cost of great effort and strenuous application. Yet it was not I, but the grace of God working with me. I beg of you, therefore, good reader, do not scorn this great labor of mind and His order as something worthless." Still, in the nearly modern seventeenth century, Robert Cawdrey introduced his dictionary with yet another explanation of how alphabetization works: "Nowe if the word, which thou art desirous to finde, begin with (a) then looke in the beginning of this Table, but if with (v) looke towards the end. Againe, if thy word beginne with (ca) looke in the beginning of the letter (c) but if with (cu) then looke toward the end of that letter," and so on. It was a tough concept.

Alphabetization had trouble taking root not just because it's conceptually confusing. Space, time, and atoms conspire to make it hard to alphabetize information that is not yet complete. The very first alphabetized list Daly found—the inscriptions on the isle of Kos—included blank spots so names could be inscribed later. Clerks compiling papyrus tax rolls in the first century B.C.E. in Egypt had to guess how much blank space to leave for each person, resulting in some entries being overcrowded and others having lots of white space. Seventeen hundred years later, the editors of the great French *Encyclopédie* faced another limitation caused by the intersection of alphabetization and physics. Because they released the volumes in alphabetical order over the course of twenty-seven years (1751–1778), the editors had to plan at the beginning not only every topic but every cross reference: They wouldn't want to add a last-decade entry on, say, sharks if the long-published volume with the carnivores article lacked a "See also" note.

These problems go away when each item has its own paper card that can be shuffled into the mix as required. Nowadays, you can get all the cards you want down at the local stationery store, but until paper became relatively cheap, in the fifteenth century, such slips would have been an extravagant use of expensive parchment. As Daly points out, there wasn't even a word for *slip* in Greek or Latin throughout the Middle Ages.

But now that paper's cheap, the old problem hasn't gone away entirely. The chances that AAAAA Towing Service was founded by five guys named Arnold, Alan, Arthur, Ashton, and Alphonse are slim.

We all understand that the company's owners made up the name so they'd get placed first in the yellow pages, because the trick works. When we have no reason to prefer one company to another, some good percentage of us are likely to call the first one listed, even if it's listed there only because it knows how to game the alphabetization system. Businesses have to play this game because traditional yellow pages suffer from the two great drawbacks of paper: They can provide only one way of organizing information, and they have so little room. When all you have to go on is a company's name—or maybe the promotional information in a display ad—you might as well just pick the first one listed. For businesses that have worked hard to provide genuine competitive advantages—faster service, better-trained employees, lower prices, maybe generations of building a good reputation in the community—getting trumped by the guys who put five *A*'s in their name is a cruel practical joke enabled by the limitations of paper and the capriciousness of alphabetical order.

THE NATURAL ORDER

Mortimer Adler, who died in 2001 at the age of ninety-nine, was once among the most prominent public intellectuals in the United States. In the 1950s he created the Great Books of the Western World series. In the 1980s, he devised a topical index to the *Encyclopaedia Britannica*. When he reflected on the course of his intellectual career, he had one clear enemy: alphabetical order.

Adler was not alone in his distrust of what he called "alphabetiasis." Theologians of the day railed against the fact that the French *Encyclopédie* was arranged alphabetically, because it demeaned God's order. And the theologians were right: The encyclopedists knew that had the *Encyclopédie* been arranged by topic, the disdain with which they held theology would have been obvious by its placement. In the next century, the poet Samuel Taylor Coleridge wrote:

> By the bye, what a strange abuse has been made of the word encyclopaedia! . . . To call a huge unconnected miscellany of the "omne

scibile", in an arrangement determined by the accident of initial letters, an encyclopaedia, is the impudent ignorance of your Presbyterian bookmakers. Good night!

Coleridge's plan for his own *Encyclopedia Metropolitana* eschewed alphabetization, organizing topics into five major classes: pure sciences, mixed sciences, history, geography and biography, and miscellaneous. He failed to complete it.

But few committed their lives to fighting alphabetization as wholeheartedly as did Mortimer Jerome Adler. The son of an immigrant jewelry salesman, Adler dropped out of school at fourteen to go to work for the *New York Sun*. He took night courses at Columbia University, got hooked on Plato, and enrolled as a philosophy major. He became so absorbed in his studies that he forgot to take physical education and wasn't graduated; Columbia nevertheless hired him as an instructor and, a few years later, awarded him a doctorate. There he participated in the Honors Program, a course of study that focused on reading the classics, Adler's passion. Throughout his life he tried to infuse the public with this love, founding the Institute for Philosophical Research and the prestigious Aspen Institute, and serving on the board of the Ford Foundation and the *Encyclopaedia Britannica*. His life yielded not one but two autobiographies.

Adler is perhaps best remembered for three major works, including *The Paideia Program,* an attempt at an ideal educational syllabus. The other two attempted to map knowledge in conspicuously non-alphabetical ways.

In 1952, the *Encyclopaedia Britannica*'s parent company published his Great Books of the Western World, a chronologically ordered set of 443 of the great works of Western history, in fifty-four volumes. The first two volumes were Adler's hugely ambitious *Syntopicon,* a listing of 102 Great Ideas, as well as an Inventory of Terms that Dwight Macdonald, in a scathing 1952 review, described as "1,690 ideas found to be respectable but not Great." "He has the classifying mind, which is invaluable for writing a natural history or collecting stamps," said Macdonald. The *Syntopicon* cost half of the $2 million the Britannica company spent on the Great Books series and occupied much more than half of the eight years Adler and his team spent on the series.

Adler brought his antipathy toward alphabetization with him when he joined the board of the *Encyclopaedia Britannica.* He noted that between 1949 and 1974, "the most insistent and vexatious problem discussed at board meetings was the choice between an alphabetical and a topical organization for the next edition." (One imagines that this was a problem only because a certain board member was insistent and vexatious.) At last, the chairman of the board, Senator William Benton, settled the issue: Since no topically arranged encyclopedias had succeeded, the *Britannica* would stay alphabetical. Adler kept at it. Yet after he became the chairman of the board of editors, he still couldn't get the idea through. At last the *Britannica* agreed to fund a project that would offer a topic-based alternative. Eight years later, the *Propaedia,* Adler's Outline of Knowledge, was ready, with 186 sections clustered into ten topics and published in 1974 as part of the fifteenth edition of the *Britannica.* "The whole . . . deserves to be read carefully," he said, a task the Propaedia's length and dryness makes unlikely.

Adler fought Ahab-like against the alphabetical because he was sure that "inherent in the things to be learned we should be able to find inner connections." "Resorting solely to the alphabet" is "intellectual dereliction," "an evasion of intellectual responsibility," and an "intellectual defect," he wrote in *A Guidebook to Learning: For a Lifelong Pursuit of Wisdom*—and that's just within a page and a half. Adler understood that reasonable people might disagree about how exactly ideas connect—hence he crowed that the *Propaedia* "captures the intellectual heterodoxy of our time"—but that only led him to grant that the ten big topics he'd discerned should be arrayed as a circle, rather than ranking some as more important than others. Readers got to shuffle the cards, but Mortimer Adler was confident that the cards he dealt reflected the basic division of ideas.

The great joke is, of course, that Adler's projects already feel hopelessly outdated. From the selection of the Great Books to the 102

Great Ideas to the confident way the *Propaedia* divides and links topics, it all seems so clearly rooted in one man's vision of knowledge. Adler himself had no regrets. Later in life he said he would have made only three changes to the Great Books—to add Apollonius's *Conics* and Fielding's *Tom Jones,* and delete *Candide*—less than a 3 percent change, as he proudly noted. But would we today issue a list of the Great Books of the Western World without a single black writer? Are we sure that Henri Bergson's work belongs on a shelf next to Plato and Wittgenstein's? Melville but not Hawthorne? Balzac but not Flaubert? "A Rose for Emily" as Faulkner's sole work? Jane Austen, Virginia Woolf, Willa Cather, and George Eliot as the only women represented? Even the layout of the book demonstrates that we are being lectured by someone who expects us to just listen: It's printed on paper so thin that it would show our notes through the other side . . . if room had been left in the margins for us to write notes. Adler's anthology of Great Books treats us like couch potatoes.

Adler's works glue together ideas based on his decisions about how they relate rather than the "disastrous" neutrality of alphabetization. You can't publish a book without using glue: The pages have to go in one order and not another, so the Mortimer Adlers of the world have to come up with an order for them. The task and the discipline are imposed by the physical limitations of paper.

In the third order of order, though, ideas come unglued. Adler's learned way of organizing the great books is of value, but other scholars shelve them differently, as may anyone who enters a bookstore to browse. In the digital order, all shelvings are provisional. As Joseph J. Esposito, the president of the Encyclopaedia Britannica Publishing Group, said in 1993, "We do not scorn chronology or alphabetization; but these ways of ordering events and ideas no longer seem so incontrovertible, so natural." Instead, he said the *Britannica* saw itself becoming more "atomistic" in the sense of being composed of small units of information that can be electronically retrieved and organized. He added, "We wonder what that says about knowledge itself."

Precisely. Alphabetical order isn't arbitrary enough, and not just because it means that AAAAA Towing gets the most calls and kids named Zywitz get their snacks last. Beyond alphabetical order is the purely miscellaneous: Every idea is browsable and ideas are instantly assembled into Propaedias and Syntopicons relevant to each person's particular needs and way of thinking. This is the world the digital order is creating.

THE JOINTS OF NATURE

The antialphabeticists have a long pedigree. Plato in the *Phaedrus* talks about reality having natural "joints" and compares knowing the world to butchering an animal: A skilled thinker, like someone skilled at carving the drumsticks off a turkey, has to know where the joints are. Arbitrary organizational schemes such as alphabetization make a virtue out of ignoring the joints. But our categorizations of animals into species, species into races, animals into sexes, heavenly bodies into planets, and atoms into elements reflect real, existing joints in nature, don't they?

This isn't an idle question with an obvious answer. The philosopher Ian Hacking, in *The Social Construction of What?,* notes that in the past forty years there has been an explosion of books and articles arguing that nature, knowledge, illness, gender, facts, emotion, quarks, and, yes, even reality are inventions, arbitrary ways of carving up the turkey. Hacking says that authors use the phrase "social construction of" to pick a fight. With the slap across the face comes the challenge that the field is nothing more than a set of lines drawn to maintain the power of the existing elite.

The social constructivists have a point. How we draw lines can have dramatic effects on who has power and who does not. Before South African apartheid ended, in the early 1990s, a fifty-page passbook that all nonwhites were required to carry at all times determined where citizens were allowed to go, what they could learn, and whom they were permitted to kiss—as specified by three thousand pages of racial laws. The lines were complex, covering many subdivisions of their four major racial categories (European, Asiatic, mixed, and Bantu), and included tests as "rigorous" as seeing if a comb could go through your hair easily. In one famous case, Vic Wilkinson, a jazz musician, was reclassified five times before he was fifty, each time with serious consequences; once he was forcibly separated from his wife and children.

But the arbitrary power of drawing lines isn't held exclusively by nations. In 1972, Dr. John E. Fryer, a psychiatrist with a medical degree from Vanderbilt University, put on a Nixon mask, a wig, and clothes several sizes too large and reluctantly addressed the American Psychiatric Association (APA) through a voice-distorting microphone as "Dr. H. Anonymous" of the "GayPA." Being gay was a "mental disorder" according to the bible of the profession, the yellow-jacketed *Diagnostic and Statistical Manual of Mental Disorders.* Fryer had to hide his identity because those with mental disorders were not allowed to be psychiatrists.

The *DSM* not only brings order to the messy field of psychiatric illness, it provides the code numbers by which psychiatrists identify diseases to insurance companies. If the syndrome isn't in the *DSM,* there's not going to be any reimbursement. The first edition, published in 1952, had only about sixty different disorders and listed homosexuality as a "sociopathic personality disturbance." In the second edition, in 1968, homosexuality was the first entry under disorder 302, "Sexual Deviations." It had become controversial enough for the APA to put together a panel discussion at its 1972 meeting, including a presentation by Ronald Gold, a forty-one-year-old gay activist, titled "Stop It, You're Making Me Sick!" Gold described a grim life path, starting with psychiatrists who tried to cure him as a teenager by shooting him with Sodium Pentothal. The speech led him to an emotional personal encounter with the head of the APA's Committee on Nomenclature, responsible for the care and maintenance of the *DSM,* who that night drafted "the resolution that a year later officially took homosexuality off the psychiatric sick list." Instead of it being a disorder classified under sociopathology—a diseased way of being with others—homosexuality was considered a problem only when the person was unhappy with her or his sexuality. In the latest edition of the *DSM* there is no entry on homosexuality at all. The APA—through a power conferred by its control of categorizations—now considers it unethical to treat homosexuality as a disorder to be cured.

Social constructivism is right that we sometimes draw lines arbitrarily, that drawing lines has real consequences, and that elites use arbitrary lines to stay in power. It doesn't have to be right in all its claimed applications for it to pose a deep challenge in an argument our culture has been having with itself for millennia. Western history began with the ancient Greek belief that not only must the world have joints, but if knowledge is to exist, humans have to be capable of discerning them. Knowledge is what happens when the joints of our ideas are the same as the joints of nature. Further, order is beauty, thought the Greeks, so knowledge and the knowledgeable mind must be beautiful as well. This conjunction of ideas shaped Western science, education, art, and government. And it led directly to one of the most remarkable ideas in history.

THE ORDER OF HEAVEN

Two thousand years after Pythagoras first came up with the idea of the harmony of the spheres, John Milton offered his kudos:

> If our hearts were as pure, as chaste, as snowy as Pythagoras' was, our ears would resound and be filled with that supremely lovely music of the wheeling stars. Then indeed all things would seem to return to the age of gold. Then we should be immune to pain, and we should enjoy the blessing of a peace that the gods themselves might envy.

The Greeks assumed that the cosmos is perfectly ordered and arranged; the word *cosmos* itself means both "all that is" and "beauty." Pythagoras therefore figured that the distance between the planets must reflect the order and harmony of the universe. But harmony is based on mathematics: Divide a string into the ratios 2:1, 3:2, 4:3, or 5:4, pluck it, and you hear something beautiful. So, Pythagoras reasoned, the heavenly spheres must fall into those ratios. Since they move, they must also make sound as they whir, a sound that must therefore be harmonious and beautiful. We're not aware of the sound because we've been hearing it since birth. It's become background "noise." Thus did the Greeks deduce that we must all live within an unheard beauty.

When Christianity came around, it added its own wrinkle. God is perfect, so when He ordered the universe, He didn't leave any holes in it. If He had, then it would be possible to imagine some more perfect God who filled in the gaps. Thus was born the Great Chain of Being, the idea that the things of the universe form a perfectly ordered ladder, with no missing rungs, from God to angels to humans to mammals to birds to insects to clams to plants to minerals to pure nothingness. Each and every thing has its place, depending on how much spirit it contains, as opposed to mere matter. For centuries, the pursuit of knowledge entailed working out the details. Not only were rabbits ahead of fish and gold ahead of lead, but squires were above merchants. Above all, the idea of perfection drove the chain. In a perfect world, if a creature became extinct, there would be an imperfect gap in the chain. Therefore, creatures can't become extinct and evolution can't happen.

Even though the harmony of the spheres and the Great Chain have fallen out of favor, we still believe there is an order to nature waiting to be discovered. The physical world isn't arranged arbitrarily, like the letters of the alphabet, nor is it based upon the whimsy of any single scholar. Science is all about finding the joints of nature. For example, no one disputes the order of the planets.

Or so it seemed until the summer of 2006, when what every schoolchild knows came unglued in public, seemingly all at once. The controversy had been brewing for years. Even while we were continuing to teach our kids "My very excellent mother just sent us nine pizzas" as a mnemonic for the names of the nine planets in order from the sun, astronomers were deep in debate not just about which objects are planets but what it means to be a planet at all. The controversy had grown more heated the year before when Caltech scientist Michael Brown identified an object in the Kuiper Belt, a group of icy bodies out beyond Neptune, about ten billion of which are larger than one mile across. This particular body, which Brown nicknamed Xena in honor of TV's warrior princess, had first been photographed in 2003, but it's so far away—a billion miles past Pluto, three times the distance of Pluto from the sun—that its motion wasn't detected until the data was reanalyzed a year later. By measuring how much light its frozen methane surface reflects, Brown was able to come up with an estimate of its size: It seems to be about a quarter the size of the earth and one and a half times larger than Pluto. Brown reasoned that if it's bigger than Pluto and it circles the sun, it should count as a planet.

Maybe. The controversy was so heated that the *New York Times* devoted an editorial to the topic:

> When a Caltech astronomer, Michael Brown, announced last year that his team had found a distant object three-fourths the size of Pluto orbiting the Sun, he declined to call it a planet, and he even suggested that Pluto should not be considered a planet either. . . .
>
> Now Dr. Brown has found something orbiting the Sun that's bigger than Pluto and even farther away. He's changed his mind and proposed that Pluto keep its designation, and that the new object, an extremely big lump of ice and rock, should also be deemed a planet. There is still no good scientific rationale for the judgment, he admitted, but this is a case where habit—75 years of calling Pluto a planet—should trump any scientific definition.

The *Times* concluded, "Our own preference is to take a cleaner way out by dropping Pluto from the planetary ranks." Why? "Scientists may well discover many more ice balls bigger than Pluto, and it's a safe bet that few in our culture want to memorize the names of 20 or more planets." Won't someone please think of the children?

It would not be the first time a planet got demoted. In 1766 a German scientist, Johann Daniel Titius, noticed that there was a mathematical relationship—shades of the harmony of the spheres—among the distances of each of the known planets (six at the time) from the sun: If the sun is taken as zero and the first planet as three, double the number and then add four to start a series that expresses the ratio of distances between the planets. In 1772, the astronomer Johann Elert Bode popularized the formula, which became known as Bode's law. He noticed that while it worked for the most part—and it continued to work when the seventh planet, Uranus, was discovered—there was a gap between Mars and Jupiter. So in 1801, when Giuseppe Piazzi discovered Ceres right where Bode's law said it should be, all seemed right with the solar system . . . until three more "planets" were discovered in the vicinity. Then more. By the end of the century, several hundred had been discovered. Piazza had found not another planet but an asteroid.

In 1999 the International Astronomical Union formed a working group with the task of coming up with a formal definition of a planet. A year before the vote was taken, Alan Stern, a planetary scientist at the Southwest Research Institute and a member of the working group, told me there were three major proposals on the table. Stern's preference: Define planets by the type of thing they are, objects of a certain size that orbit a star. Another group lobbied for defining planets by what's around them: If they're part of a swarm of objects, then they're not planets—ruling out Ceres. Third, there was "a group that thinks *planet* is a cultural term that has no business in science," Stern says. "That's really amazing to me."

Stern preferred the first definition because it's based on a real property of the thing itself. "We want a planet to orbit a star, because if it orbits another planet, it's a moon. And we want it to be the right size." But what's the right size? "That's where the controversy is," Stern says. Some had suggested adopting an arbitrary standard, such as insisting that planets have to be at least the size of Mercury (four thousand kilometers in diameter). But Stern wanted to use physics. The maximum size is easy to set: "It shouldn't be so big that it ignites in nuclear fusion like a star," he said. But how to settle the much more controversial question of minimum size? "A small object will retain whatever shape you give it because of the chemical bonds," he explained. "But if you keep adding mass, something wonderful happens: It knows that it's big. Gravity rounds it. It's an inexorable process." So, Stern suggested that the minimum size for a planet be the size at which the object becomes round. The lower limit at which bodies become round "seems to be set by nature," he concluded. In other words, Stern had found a joint in nature. Using Stern's definition, there would more likely be "nine hundred than nine" planets in our solar system. But what of the *Times'* objection that that's too many to memorize? "Schoolkids can't name all the mountains, but no one thinks mountains aren't a real classification," he counters.

In the summer of 2006, the International Astronomical Union met in Prague and made up its mind. Sort of. Some of the details were left for the 2009 meeting in Rio de Janeiro, but the new definition begins with Stern's and adds the second contender to keep the number of new planets manageable. Thus a planet is now a star-circling body large enough to be rounded by its own gravity and one that has cleared the area around it of other objects. If you're round but haven't cleared your zone, you're a dwarf planet. The zone-clearing requirement reduces Stern's nine hundred planets to a mere eight.

There is another possible winner in the dispute, though: The third position that says that *planet* is a cultural term, not a scientific one. Stern's argument against this has much to do with the social effects of giving up the term. "Every man on the street who's seen *Star Trek* can tell what a planet is," he says. "If the IAU were to announce that there's no such thing as a planet, that it's just a cultural thing, I think my colleagues in other fields and the public would break out laughing." But, Stern as a scientist properly adds, "I'm agnostic. I want the data to inform me. I want to have a classification scheme that illuminates."

The problem is that every proposed definition—including one that would have added adjectives to the planets, making Earth a "cisjovian" planet because it comes before Jupiter—illuminates something. Each takes a cut through nature that picks out some number of objects in our solar system and calls them "planets." But why bother? There's an indefinite number of such categories we could define into existence. For example, we could define "lumpettes" as all nonround objects that circle the sun and rotate counterclockwise. But we don't because lumpettes have no properties in common beyond the ones that define them. There's nothing further to say about them, just as there's nothing further to say about what all extra-large T-shirts with a grease stain on the left sleeve have in common.

Using this argument, some biologists deny that race is a scientific category. *Species* matter, the argument goes, because the differences among species affect how well they survive. Scientists can't explain evolution without talking about species. Race, on the other hand, picks out a set of properties that make no more biological difference than eye color, hair color, or whether you're left- or right-handed. Race is the "lumpette" of biology.

Of course, we can choose to divide our species into races, just as we can sort ourselves by curvature of the eyebrow, and there have been drastic social and historic consequences of racial sorting. But our reasons for choosing to sort by race arguably have nothing to do with science.

Likewise, we can choose criteria by which to define planets. But now that we've seen scientists voting on the definition of a planet, we know that those who argue that planets aren't worth defining have a point. Call Xena a planet and what have you learned beyond the fact that it fits criteria we've accepted? There are millions of objects circling our sun. The nine bodies we've called planets are interesting to us because we have thousands of years of lore about them. Maintaining a category of planets says less about the nature of our universe than about our need to imagine walking on spheres other than our own blue one. Planets are interesting not because of what they are but because of who *we* are and where our dreams are set. This doesn't mean that planets are without meaning. On the contrary. Our insistence on maintaining the category even though there is no compelling *scientific* reason to do so exposes a deeper meaning that is becoming more important as more realms break free of their categorical tethers and join the swirl of the miscellaneous: How we organize our world reflects not only the world but also our interests, our passions, our needs, our dreams.

CHEMICAL SOLITAIRE

The planetary club may have had its rules of admission jiggered to make sure it only admits the popular orbs, but the chemical club seems stricter. No one argues whether, say, lithium—discovered in 1817 by a Swedish chemist who was analyzing the mineral petalite—is an element. Likewise, no one denies that it's soft, silvery, and tarnishes rapidly. Nor does anyone argue with its placement in the standard periodic table of elements, high and to the left.

Whether Pluto is in the planet club tells us nothing interesting about Pluto, but lithium's placement in the table of elements tells us lots. Because it's immediately to the left of beryllium, we know lithium has one fewer proton than beryllium. That it's in the leftmost column tells us that it's an alkali metal, can be easily cut, and is highly reactive. That it's in the second row of the leftmost column tells us that it's the lightest of the light metals. Because it's in that row, we know how its electron shells are organized. All of these facts are disclosed in the grid's organization. The periodic table's layout therefore seems to be the opposite of an arbitrary order like alphabetization that adds no information to the items it arranges. The table has laid bare some real joints of nature. If anything should survive untouched by the third order of order and the rise of the miscellaneous, it ought to be the periodic table of the elements.

The periodic table can be traced back to the German chemist Johann Döbereiner, who, also in 1817, pointed out that various groups of elements form triads. For example, in the lithium-sodium-potassium triad, the first reacts mildly to water, the second reacts more strongly, and the third explodes. In 1862 a French geologist, Alexandre-Émile Béguyer de Chancourtois, discovered that the triads aligned vertically if he plotted the elements on an upright cylinder into which he'd etched a line rising at a forty-five-degree angle. This lined up elements whose atomic weights were sixteen numbers apart. He had discovered the organizing principle that yielded Döbereiner's triads. "The properties of the elements are the properties of numbers," Chancourtois said.

Two years later, the English scientist John Newlands took the search for the hidden order of the elements a step further and structured a chart of the elements around octaves, a harmony of the itsy-bitsy spheres, so to speak. Although he was a well-respected chemist, when he presented his ideas at a meeting of the Chemical Society in London, one of the attendees mockingly suggested that he arrange the elements alphabetically and look for patterns there.

A Russian chemistry professor, Dmitrii Ivanovich Mendeleev, had an intensely practical motive for looking for a way to order the elements compactly. He was writing a two-volume textbook, and at the end of the first chapter he discovered that he had covered only eight elements. He needed a way to treat the remaining fifty-five and their relationships in the remaining space his publisher had allotted.

Mendeleev was a freethinker. Later in his career he rode in balloons, invented a smokeless gunpowder, tried to become an Arctic explorer, helped design an important protectionist tariff, fought the Spiritualist movement, wrote art criticism, and supported women's education. He also did not believe that matter was made of discrete atoms that have their own inner structure. When electrons were discovered, in 1897, he denied their existence, even though the patterns of electrons explains relationships in the table he developed.

Unburdened by theory, Mendeleev laid out scraps of paper as if he were playing a version of solitaire, until he found a pattern that made it easy for students to remember the properties of the elements. He put elements that shared certain properties into columns, so if you knew the properties of one element, you would also know the properties of all the other elements in its group. The rows he arranged in size order. He even found meaningful relationships along the diagonals: Draw a diagonal line from boron to astatine, and the elements to the lower left of the line are metals. Mendeleev didn't know why his system worked. He just knew that its spatial arrangement of the elements expressed their properties.

Mendeleev so believed in the primacy of order that in his chart of the sixty-three known elements he left spaces—appropriately, it turned out—for three undiscovered elements with particular atomic weights. Like Bode looking for a planet where the formula of order said one ought to be, Mendeleev assumed that nature was so orderly and complete that there couldn't be blanks. How could the great chain of elements have any empty rungs? How could nature play in perfect harmony if some of the notes were missing?

Early in the twentieth century, however, Henry Moseley proved that Mendeleev was charting the wrong property. Moseley—whose young death in World War I led the British to exempt scientists from combat duty—discovered that there was indeed a property of the elements that let them be lined up in sequential, numeric order. But it wasn't atomic weight—protons plus neutrons—as Mendeleev and his colleagues believed; it was atomic number, simply the number of protons. This meant Mendeleev's chart got a few elements wrong; for example, iodine and tellurium had to be switched. But because atomic weight is roughly associated with atomic number, the switching of seats was no worse than might happen at a one-hundred-person banquet.

The relationships found in the current version of Mendeleev's periodic table are real: We can't make helium into a metal by changing its position. Even the fact that the table has been altered over the years—after William Ramsay discovered the "noble" (inert) gases in 1894, a new column was added—shows that the periodic table reflects relationships that are in the world, not just in our heads.

But Mendeleev's layout is not the only way to present those relationships. The alternatives include a triangular version developed by Emil Zmaczynski that "shows the pattern of filling electron shells"; Ed Perley's circular version, that better demonstrates "the electronic orbital structures"; and PeriodicSpiral.com's spiral version, that claims to more accurately represent "hydrogen's ambiguous relationship to the noble gases and halogens." A very popular spiral version, created in 2005 by Philip Stewart, a plant scientist, depicts the elements on top of an image of a swirling galaxy. The Royal Society of Chemistry in the United Kingdom sent a poster-sized version of Stewart's table to every secondary school in the country. "I hope my table will help by conveying the message that the matter of which we are made is the same as the stuff of the stars," he said, repeating the ancient and beautiful Greek idea that the microcosm reflects the macrocosm, that order is the same everywhere you look.

Each of these tables presents real relationships. But whether you care about those particular relationships has everything to do with how you work in the world. If none of them captures the relationships that matter to you, then you can come up with your own. For example, L. Bruce Railsback, an earth scientist at the University of Georgia in Athens, spent four years devising a periodic table that organizes the elements the way earth scientists think about them. Unlike chemists, earth scientists mainly encounter elements not in their pristine, prototypical form but as ions. He realized how poorly suited the standard table of elements was when, in 1999, he was teaching a class about the behavior of minerals in the earth's waters. "I looked like a contortionist trying to point to different elements in different places," he said. His new version clusters the ions by their positive or negative charge. He even allows elements to be listed in more than one spot, violating an unspoken premise of Mendeleev's tables. For example, sulfur shows up four times on Railsback's chart because that "reflects the different ways sulfur can behave in nature." The premise that elements have only one place in the natural order was actually built right into Mendeleev's method of constructing his table: He laid out slips of paper and assumed he needed only one slip per element. If matter limits the first-order organization, paper historically has limited the second order.

We should not conclude that all arrangements of chemical elements are equally good. An amateur might mistakenly list lithium as a noble gas. Other versions might be of no interest to scientists of any stripe. But neither should we conclude that there is only one way of organizing the elements that reflects the joints of nature. Consider why we have no periodic table of recipes.

In the 1980s, the manufacturers of the early personal computers kept pointing to the same example to prove that every household ought to have at least one: PCs would be the new cookbooks. Instead of having to figure out how to turn a recipe for four into a recipe for five—quick, what's five-fourths of three teaspoons of vanilla?—the computer would calculate the amounts instantly. And if you had left-over tarragon, potatoes, and sliced ham, the computer in your kitchen would find all the recipes that use those ingredients.

Most of us still don't have computers in our kitchens, but the PC manufacturers had a point. Paper-based cookbooks don't think about food the way we do. They don't know that we have some dill on hand but not tarragon, that our carb-counting spouse won't eat potatoes, and that two of our three teenagers decided last week to become vegetarians. Cookbooks don't know that our family considers pizza to be a breakfast food and pancakes a fine dessert. They don't know that we bought twice as much fresh broccoli on sale than any human could be expected to consume. How could they know any of this? A cookbook is printed on paper, unchanging throughout its lifetime, and is the same for everyone who buys a copy . . . at least until we start writing notes in the margins, drawing X's through the recipes that don't work, and dog-earing the pages of recipes that do.

We're okay with the failings of printed cookbooks because we understand the limitations of paper. Besides, the fact that they list pancakes as a breakfast food instead of as a dessert is at most an inconvenience, not a mistake in the natural order. (And when you know what you're looking for, there's always the index.) Nevertheless, if we could make our own cookbook, it would not only reflect our own preferences in food, it would be different every time we opened it. If we'd just bought broccoli, it would offer every broccoli recipe on the first page—it'd be more like the Web site Epicurious.com than like a bound book—and when we browsed for breakfast foods, it'd suggest last night's pizza leftovers. When a friend who's allergic to wheat is coming to dinner, we want to slice our foods differently. Not only don't we want a single way of arranging our recipes, we don't even want multiple ways. What we really want is to miscellanize our cookbook so that every ingredient and recipe can be combined with any other based on our permanent tastes and momentary situation. Such a cookbook reflects the miscellaneousness of our needs, preferences, and refrigerator contents. A periodic table of recipes would just get in our way.

There is a difference, of course, between Julia Child and Dimitrii Mendeleev. If cookbooks disagree about whether baked stuffed potatoes are a side dish or an entrée, we don't think anything is at stake other than the author's taste. But a table of the elements that classifies hydrogen as heavier than lead is just plain wrong. As Umberto Eco says, there are many ways to carve a cow but none of them include serving a segment that features the snout connected to the tail. Yet because of the limits of second-order media, such as paper, we've had to pick some orderings over others, a limit the third order of order removes. Now we know that not everything has its place. Everything has its places—the joints at which we choose to bend nature.

THE GEOGRAPHY OF KNOWLEDGE

From across Fortieth Street all seems well with the New York Public Library's mid-Manhattan branch. Although it's not the largest—the one with the lions in front holds that honor—since it opened in 1970 this has been the branch to visit if you want the best chance of borrowing a best seller. The library also makes available, on its third floor, over one million graphic arts prints. On the fourth floor, city residents use computers to access the Internet for free. On the sixth floor, they can attend frequent, free lectures. It's a terrific library. But it's built upside down.

Melvil Dewey would notice that immediately. When, in 1876, he published the library classification system that bears his name, he carefully assigned books about philosophy the lowest range of numbers (the 100s) because, of course, philosophy laid the foundation for all other studies. Next came religion, the 200s, which to Dewey gave truth its content. Then came the social sciences (the 300s), followed by language, natural sciences, and math (the 500s), technology and applied sciences (the 600s), arts and recreation (the 700s), literature and rhetoric (the 800s), and finally geography, history, and biography (the 900s). Dewey came to think that the physical layout of libraries should reflect this basic structure of knowledge. But the mid-Manhattan library gets it all wrong. Rather than books on philosophy and religion being shelved on the first floor, they sit on the fifth-floor shelves. Worse, the corner of the fifth floor that faces Fortieth

Street mixes the foundations of Dewey's order of all knowledge—philosophy and religion—with mere biographies, and history shares its shelves with the social sciences. Mr. Dewey, were he around, would not be amused. He would take comfort, though, that even if the mid-Manhattan branch of the New York Public Library system has not hewn strictly to the lines he drew, it at least has a geography.

Ninety-five percent of public school libraries in the United States and 200,000 libraries worldwide use the Dewey Decimal system. Yet in the right company, librarians frequently react to a mention of the Dewey Decimal system with a roll of the eyes and an apology about it being out of date, provincial, even embarrassing. In 2005, technologists had a good laugh at its expense at the Emerging Technology conference in San Diego. At a talk on the future direction of digital "information architecture"—the art and science of organizing electronic information—Clay Shirky, a professor at New York University, addressed an audience of techies, many of whom consider the Web to be a second home. Shirky put up a slide showing that eight of the nine major divisions under the religion classification are explicitly for Christian books. Dewey's organization of religion just gets worse the closer you look at it. Judaism occupies its own whole number (296), but Islam shares its number with two others, Babism and Baha'i (297), even though many Muslims consider Baha'i and Babism to be apostate, Johnny-come-lately cults. At least Muslims got placed at the top level of the system, unlike Buddhists; as a subcategory of "Religions of Indic Origin" (294), Buddhism falls to the right of the decimal point.

Religion is not the only problematic category. Even the title of the 100s, "Philosophy and Psychology," raises red flags. These days, philosophers think they are doing something broader and deeper than psychology, while psychologists think they're doing something more focused and useful than philosophers. The psychologists and philosophers would unite in opposition, however, to giving over the entire 130s to "Paranormal phenomena," with subtopics such as "Occult methods for achieving well-being" (131), "Divinatory graphology" (predicting the future by analyzing handwriting; 137), and

"Phrenology" (discerning personality by examining the bumpiness of the head; 139). Dewey's system puts phrenology on a par with Aristotle (185) and Oriental philosophy (181).

Dewey's arrangement of the top-level categories has gotten less appropriate over time. The speakers of "Ural-Altaic, Paleosiberian, and Dravidian" get their own whole-number category (494) but the 1.2 billion who speak Chinese do not. And there's still a special category for "Education of women" (376), dating back to when educated women were a special case.

These anomalies—and occasional insults to entire religions—come about in part because the Dewey Decimal Classification system reflects Dewey's perception of the topical distribution of books in 1876. If the Buddhists wanted to make it into the list of a thousand top-level topics—the ones designated by whole numbers—they should have written more books in the nineteenth century, or at least gotten a better PR agent. Or maybe not. There's no indication that Dewey actually surveyed American (much less world) libraries to get a factual basis for his classifications.

So why don't the people who run the Dewey Decimal system fix it? They're not small-minded American Christian jingoists. They're librarians who understand that Dewey's original schema is embarrassing in the modern era. If we want to see how the physical world has silently shaped how we put together our ideas about the world—and why any traditional classification scheme is bound to embarrass somebody—there is no better example than the Dewey Decimal system.

DEWEY'S WORLD

When he was fifteen, Dewey bought a pair of cuff links inscribed with an *R*, which, he wrote in his diary, were to be "a constant reminder . . . that I was to give my life to reforming certain mistakes and abuses." In his time he led movements not only to organize libraries but also to simplify spelling, popularize shorthand, and switch to the metric system. All of these efforts would use standardization to drive out inefficiency, whether it was the time wasted writing "through" instead of "thru" or the energy expended remembering if quarts have sixteen or thirty-two ounces. He quickly set up two committees to explore the positive effects of these reforms, which, he later wrote, reported "unanimusli" that if "we had scientific spelling and abolisht the absurdities of 'compound numbers' and used only international weits and measures," children would have saved three to four years by the time they graduated from college. A true believer, Dewey simplified the spelling of his own name to Melvil Dui in 1879. When he was appointed Columbia University's first chief librarian, he returned to spelling his last name Dewey, but stuck with Melvil.

This great believer in the power of rationality had humble beginnings. He was born in December 1851 to shopkeepers in Adams Center, a small town in western New York, population five hundred. At eighteen, he entered Amherst College, then an orthodox Christian school that built good Protestant character by teaching the classics. To pay off his debt, while still a student Dewey took a job keeping the account books at the college library. Even though the library wasn't used much at Amherst—students were expected to stick close to the books the professors assigned—Dewey saw that libraries could refashion education if they threw open their shelves to ordinary citizens rather than requiring them to go through librarians. Libraries could be more than repositories. They could empower every individual to become a lifelong learner.

But opening the shelves wouldn't do much for education if the ordinary citizen couldn't find a book, or even know what books were there. The organization of the books had to convey information about them, or the library would remain a mere warehouse of random titles, typically ordered alphabetically by the authors' names, and sometimes by the size of the book. Librarians located books for patrons by consulting a bound catalog, which also served to track inventory. Starting with the poet Callimachus, who compiled a 120-volume catalog of the more than 400,000 scrolls in the library at Alexandria in the third century B.C.E., catalogs frequently used some element of topical categorization. Callimachus divided Alexandria's holdings by type of author—poets, lawmakers, historians, etc.,—within which scrolls were listed alphabetically by author, further divided by era, format, and subject. In the Renaissance, the increased number of books—fifteen to twenty million books were printed in just the fifty years after Gutenberg invented the printing press—led many libraries to arrange their books by size in order to cram everything into the available shelf space. Retrieving information from a library required an expert.

Dewey's plan to democratize libraries, and thus to democratize knowledge, pulled together three big ideas, each of which was in the air when he started to work on his plan as a twenty-one-year-old student at Amherst. Dewey's genius was in synthesizing ideas, though it didn't hurt that he turned out to be a fearsome organizer of associations to promulgate them.

The first idea was simply that there could be a single, universal way of cataloging books, one that all libraries would use. Dewey was developing an idea implemented by the British Museum's Department of Printed Books, where, in 1848 and 1849, the Scottish historian Thomas Carlyle, famous for his passionate and heroic histories, squared off in an appropriately epic battle over cataloging.

Perhaps it was just that the British Museum made Carlyle cranky. He complained that the room was noisy, crowded, infested with lice, and that one fellow there blew his nose every half hour. Worse, people in the room were engaged in such trivializing tasks as writing encyclopedias and biographical dictionaries. The circumstances were so stressful that, Carlyle said, every time he entered the reading room he got a "Museum headache." One person irked him particularly: Antonio Panizzi, in charge of the museum's library. Panizzi's great offense was not permitting Carlyle to access the quiet inner rooms where King George III's library was kept. So when the opportunity arose to contest Panizzi's competence—"Vulture Panizzi," as Carlyle called him—the battle was joined. Panizzi was no stereotypically timid librarian: A death sentence awaited him in Italy for his role in the effort to unify the country.

Panizzi, an advocate for opening libraries to the common folk, had devised ninety-one rules to guide the creation of a new, friendlier catalog of books, calling for long, consistent entries that aggregated information uniformly about each edition and individual copy. Carlyle argued that Panizzi's catalog would take too long to produce. Carlyle preferred to have an incomplete and imperfect catalog sooner. But he also lobbied for additional volumes that would cluster books by topic, for a very personal reason: A few years earlier he had heard that the museum had a collection of books about the French Revolution, but because he didn't know who'd written them, he couldn't find them. Carlyle lamented, "For all practical purposes this Collection of ours might as well have been locked up in water-tight chests and sunk on the Dogger-bank, as put into the British Museum." In the end, the commission investigating the issue of catalogs agreed with Panizzi's insistence that having a uniform system would lead to long-term benefits. Not all of Carlyle's supporters took it well. Said one: "The fat pedant and Italian language-master proved more than a match for the Scottish man of genius."

Panizzi's rules likely inspired Charles Coffin Jewett, a Smithsonian librarian who in 1852 wrote a set of his own rules for building a library catalog. He realized that if libraries expressed book information in a standard way, it would be much easier for them to share information. Further, he thought, you couldn't merge the catalogs for two libraries if the catalogs were themselves bound ledgers. So Jewett came up with the idea of printing the entries on cards.

In 1873 Dewey read an article by Jewett and was inspired to come up with a standard way for libraries to organize the books themselves. Dewey had also read an 1870 article in the *Journal of Speculative Philosophy* that recommended organizing books alphabetically within subjects. This required arranging books relative to one another rather than fixing them to spots on shelves, as was the current practice. Dewey noted, "Of this I am inclined to be a friend."

But what should such a scheme look like? Here was Dewey's second big idea. The alphabetical ordering of books that was typical of libraries at that time worked when you knew exactly what you were looking for, but not if you didn't know what was available or had only a narrow or vague understanding of a particular topic—deficiencies that could befall even a Scottish man of genius. So, Dewey thought, why not arrange books by subject? That way, you could look up an alphabetical listing of topics and go straight to the part of the library that housed the relevant books.

It seems like a simple idea, but it required getting past the concept of a library as a warehouse with books assigned spots on shelves, just as inventory is shelved in a manufacturer's warehouse. The floor plan of the library would be a map of ideas.

Dewey was not the first person to spatialize ideas. The ancient notion of a "memory palace," a mnemonic device that has us place what we wish to remember in rooms of a building we've imagined, dates back at least to a grisly myth recounted by Cicero in the first century B.C.E., according to which the ancient Greek poet Simonides of Ceos recalled every victim in a building that the gods had crushed by remembering exactly where they had been. In the modern world, the most prominent practitioner of the technique is the fictitious serial killer and cannibal Hannibal Lecter, whose phenomenal memory (we're told) relies upon his construction of a magnificent memory palace. Because the ideas Dewey was organizing were housed in physical books, his "memory palace," inevitably, organized physical space.

A memory palace is an entirely arbitrary and personal way of arranging ideas. Dewey wanted an organizing scheme that expressed the actual relationships among them. Only where does a twenty-one-year-old student come up with a map of all knowledge? It helps, of course, to suffer from the arrogance of youth. But Dewey did not simply sit down with a blank piece of paper. Dewey acknowledged a debt to Sir Francis Bacon, who in 1623 had divided knowledge into three parts—history, poesy, and philosophy—which, according to Bacon, reflected the three capabilities of the mind: memory, imagination, and reason. The most direct influence, however, came from the early-nineteenth-century German philosopher Georg Wilhelm Friedrich Hegel, who sniped that "Bacon's philosophy is like that of 'shopkeepers and workmen.'" Philosophers don't get much cattier than that. A greater philosopher, Hegel suggested—and let's pause a moment to consider whom he might have had in mind—would get past the mere natural and experiential. So Hegel reversed the order of knowledge Bacon had proposed, putting philosophy first. Amherst College was headed by a Hegelian when Dewey was there, and Dewey adopted his reasoning.

Dewey had found his big-boned joints of knowledge. To slice it up more finely, he enlisted Amherst faculty members. One can imagine their bemused skepticism when the young man who had been their student the previous year now told them that he was creating an organized list of all possible topics of knowledge. Several took him up on it, and two of them, including his philosophy professor, Julius Seelye, came regularly to the Amherst College library to work on the task. Dewey's diaries record no disagreements with the faculty members' recommendations. He seems also to have taken his college textbooks as a guide. For example, Dewey's division of physics almost exactly repeats that in the science textbook he used in his junior year. As Dewey's biographer Wayne A. Wiegand writes, the organization of knowledge Dewey produced solidified "a worldview and knowledge structure taught on the Amherst College campus between 1870 and 1875"—a worldview and structure that assumed that the West was the most advanced culture and that Christianity laid the foundation of truth.

In March 1873, when he was still an undergraduate, Dewey had his third big idea, inspired by an 1856 pamphlet titled "A Decimal System for the Arrangement and Administration of Libraries," written by Nathaniel Shurtleff, who worked at the Boston Public Library. As Dewey wrote at the time, "My heart is open to anything that's either decimal or about libraries." In fact, fifty years later, Dewey would attribute the idea to order topics by decimal numbers to an epiphany during a Sunday sermon. Dewey was already infatuated with decimals. He wrote a school essay on the metric system when he was sixteen. When he was twenty-five he founded the American Metric Bureau to lobby for the adoption of the metric system within the United States. He even arranged his travel so that he would arrive on the tenth, twentieth, or thirtieth day of the month . . . rationalism crossing over into superstition.

Decimals offered Dewey an infinity of subdivisions; by placing topics to the right of the decimal point, he could stretch his subject areas without limit. But decimals brought serious disadvantages as well. Dewey had to hack and hew knowledge into a thousand topics—ten top-level classes each with ten divisions (although Dewey left some empty spots in the 000s, "Generalities"), each with ten sections—not because that's how knowledge shaped itself or how books sorted themselves, but because Dewey loved decimals.

To see the difficulty this presents, imagine that you are organizing Melvil Dewey's new kitchen. He is so besotted with decimals that he insists that there be exactly ten cabinets, each with ten shelves, and each shelf with exactly ten spots. You are required to fill each of those thousand spots with exactly one item. You might not have trouble deciding the sorts of items each of the ten cabinets will contain: dishes and glasses, cutlery, condiments, breads, cans, beverages, baking goods, boxed goods, bagged goods, and empty containers for leftovers. But as you try to find exactly ten types of shelves and exactly ten types of items to place on them, it gets harder and harder. What are the chances that you're going to have exactly ten spices to go into the ten spots allocated for spices in the condiment cabinet? Suppose you only have seven. Would you classify pickle relish as a spice? Might you decide that you should count Dewey's bottle of three-spice Chinese flavoring as three? When you're finished with the spices, you'll have the same challenge filling the nine other shelves in the cabinet. When you've finally unpacked every carton, you'll probably find that your organizational system has turned a melon baller into a bona fide piece of silverware, on a par with forks and spoons, and that you're now counting those chocolate sprinkles as a spice after all.

There's another problem with Dewey's use of decimals: Numbering systems have an implied hierarchy. Lists conventionally put the most important items first—No. 1 is the big punch line of a David Letterman top-ten list. Further, when we were taught how to write an outline, we learned that item III is of broad importance and item III.A.3.c.iii.0057 is a detail. So even though going to the right of the decimal provides Dewey's system with an infinite amount of room, there is inherent importance in being a top-level category toward the top of the list and to the left of the decimal point. Furthermore, if you have only a fixed number of top-level categories, what do you do when something new and important comes along?

In the 1980s, the editors of the Dewey Decimal Classification system had to decide where to put the burgeoning field of computer science, a topic that Melvil Dewey could not have envisioned. Although it seems that computer science should go in the 600s, with other "Technology and Applied Sciences," the editors instead stuck it into 000, "Generalities," where Dewey had put bibliographies, encyclopedias, and other general works that didn't have a clear place. Why? Because the 600s were filled and there were some unused numbers in the 000s; computer science got 004 (data processing and computing science), 005 (computer programming, programs, and data), and 006 (special computer methods). That way, existing sciences in the 600s didn't have to be shoved aside. Then, in the twenty-second edition of the system, in 2003, computer science achieved the dream of all ambitious topics: It was promoted to the highest level in the system. The 000 class was renamed "Computer science, information, general works." Perhaps that makes sense or perhaps in fifty years the promotion will seem to be a silly overestimation of the importance of computer science. Today's category easily becomes tomorrow's embarrassment.

If the editors can make room for computer science and even elevate its status, why don't they demote phrenology or promote Buddhism to a whole number? For all of the work of its dedicated editors, the Dewey Decimal Classification system remains weirdly out of date, reflecting the small-town sensibility of a student at a tiny Christian college in the mid-1870s. In defense, Joan Mitchell, the system's editor in chief, in her unpretentious work space at the Library of Congress, points out that changes are made every week. She refers to major work vacating some Christian topics in the 200s to make room for a more diverse mix. "There are also subtle changes in terminology," she says. "We turned 'Children born out of wedlock' to 'Children of unmarried parents.' We're always doing that sort of stuff. 'Cohabitation' used to be under 'Sexual relations.' We moved it to 'Types of marriage and relationships.' " When I ask her the number, she responds, "It's 306.841," checking it quickly.

Still, phrenology is a whole-number topic and Buddhism isn't. Why not just fix it?

Because it can't be fixed. The Dewey Decimal system is caught in a problem endemic to large classification systems tied to the physical world. Imagine that the system's editors decide to fix the system once and for all. They move discredited categories such as phrenology to the right of the decimal place. They consolidate the Christian topics, pull Buddhism up to a couple of integers, push Baha'i down, drag computer science into technology, demote philosophy from its top-ten status because, frankly, philosophy isn't the queen of the sciences anymore, and do a thorough housecleaning. What happens next?

Tens of thousands of librarians around the world pick up their razor blades and scrape the white numbers off the spines of millions of books, muttering under their breath about those damn editors who don't understand that every little change means that librarians inhale toxic white dust. Entire card catalogs get discarded, so to speak, and millions of new cards printed up. Books are piled up, moved from this shelf to that. And at the end of the months or years of work, the complaints start. The Sunnis and the Shiites are upset because they've been put at the same level. The Jews are furious because the Jews for Jesus, whom they view as Christian predators, are listed under Judaism. Feminists and fundamentalist Christians find themselves making common cause to get studies of pornography removed from the fine arts section. East Somewhere is furious because it doesn't recognize West Somewhere as a legitimate country. Librarians are out buying razor blades in bulk and white ink by the gallon.

There is no end to it. The Dewey Decimal Classification system can't be fixed because knowledge itself is unfixed. Knowledge is diverse, changing, imbued with the cultural values of the moment. The world is too diverse for any single classification system to work for everyone in every culture at every time.

But that's not a good enough answer if you're organizing physical objects. The new book that's come in has to go on some shelf somewhere. Just as Staples can't stock cables in every spot in the store, libraries have to make decisions about where to put each volume. Card catalogs do provide some flexibility—a book on military music might sit on a shelf with military books but be filed in the card catalog under both "Military" and "Music." Why not assign multiple numbers? Mitchell says that's the price of designing for libraries that put the books directly in readers' hands. In Europe, she says, more libraries have closed stacks, so they are willing to classify books under multiple categories. If you've closed the stacks and rely on librarians to fetch the books, it doesn't matter how they are physically arranged; the librarian just needs to know which shelf in the "warehouse" to go to.

And there's the weakness and the greatness of Dewey's system. The Dewey Decimal Classification system lets patrons stroll through the collected works of What We Know—our collective memory palace—exploring knowledge the way they explore a new city. But the price for ordering knowledge in the physical world is having to make either-or decisions—ironically, a characteristic more often associated with the binary nature of the digital world. The military music book is shelved with the military books or with the music books, but it can't go with both. The library's geography of knowledge can have one shape but no other. That's not a law of knowledge. It's a law of physical geography.

CARNIVAL AMAZON

If the Dewey Decimal system feels like a Victorian sitting room with furniture that's too heavy to lift, Amazon's Web pages feel like the midway at a carnival where every inch of ground is given to attracting your attention. When it comes to categorization, Amazon doesn't care about the precision and orderliness of its system; it cares about putting information—and offers—in front of you. Rather than annoying you with irrelevant ads, Amazon populates its book pages with an information-intense display of ways for a potential customer to stumble across books she didn't know she was interested in. If that means pulling together books from multiple shelves and violating Dewey Decimal categories, Amazon doesn't hesitate.

This leads to a very different user experience. Let's say you want a copy of *The Little House Cookbook: Frontier Foods from Laura Ingalls Wilder's Classic Stories,* by Barbara M. Walker. If you look up the title at the New York Public Library, you'll find fifty-two copies across the many branches. Most put it in the children's room, but the Donnell Library Center puts it in the reading room. Every one of those branches, however, has it listed under its call number: 641.59 W. That translates to:

> Technology and Applied Sciences > Home economics and family living > Food and drink.

That's one logical place for it. But just one.

If you search for the same book at Amazon, you'll find a similar classification scheme. But Amazon lists *The Little House Cookbook* under three categories:

> Children's Books > Authors & Illustrators, A–Z > (W) > Williams, Garth
> Children's Books > History & Historical Fiction > United States > 1800s
> Children's Books > Sports & Activities > Cooking

That helps. But suppose you are trying to find other cookbooks from the 1800s that are for adults. Amazon has the book categorized as a children's book, but immediately beneath the category listing, the site lets you check any or all of nine related categories to see what books might be slightly similar to *The Little House Cookbook.* If you want to see all books about both cooking and history without specifying that the books have to be for children or be associated with a work of literature, Amazon will happily build that list for you. It's like having a Dewey Decimal Classification system written to order.

Although it would seem as though it's figured out a better overall scheme, Amazon doesn't claim world-shaking expertise in building book classification systems. At first, says Greg Hart, director of entertainment business for Amazon, they even thought they would follow the path beaten by physical retailers, "putting the fiction books in this area, history books in that." They licensed a classification structure from outside sources and set up shop. But, notes Hart, Amazon's selection is much broader than that of any physical bookstore; he estimates the number of history subgenres alone to be in the "umpteens." "A typical physical bookstore might have 150,000 to 200,000 unique titles," he says. "We sell more unique titles than that in a month. By multiples."

To handle that volume of books Amazon had to come up with new ways of making titles accessible. "We can't just have a history area," Hart says. "It'd be hundreds of thousands of titles long and you'd never find what you're looking for." So Amazon began innovating.

"Customer reviews were deemed very, very controversial when we launched them," Hart remarks. "Publishers said you're allowing users to say that they hate a book." The response from Jeff Bezos, Amazon's founder, as Hart recalls it, was: "It will sell more books . . . just not ones customers don't like." Hart says Amazon has "millions and millions" of customer reviews. As is clear, Amazon is cagey about releasing actual numbers, but Hart states that the company has 47 million "active customers," that is, customers who maintain a free account on the Amazon site, all of them potential reviewers.

And it is in Amazon's interest to introduce you quickly to books you didn't know you wanted. Instead of relying on your knowing the ins and outs of the company's classification system, Amazon takes advantage of "planned serendipity," using collaborative filtering to come up with recommendations for returning customers. Collaborative filtering works on the assumption that if a group of people bought book A and also bought book B, others who buy A might also be interested in B. If many customers who buy *Romeo and Juliet* also buy *To Kill a Mockingbird* and *Of Mice and Men,* Amazon will recommend those two books on its *Romeo and Juliet* page. Thus, Amazon's page for *The Little House Cookbook* tells you that "customers who bought this book also bought the *Little House* novels, *My Little House Crafts Book, The Little House Guidebook, On the Way Home: The Diary of a Trip from South Dakota to Mansfield, Missouri, in 1894,* and three other books—all of which would be scattered across the shelves in a Dewey library. Amazon brings them together not because they are on the same topic but because of a statistical analysis of customers' buying patterns.

Collaborative filtering—which is far more sophisticated than this simple example shows—doesn't work perfectly. For example, the association of *Romeo and Juliet* with *To Kill a Mockingbird* and *Of Mice and Men* tells us more about which books are taught in English classes than about the tastes of those purchasing them. So Amazon not only lets users tune their filters—perhaps you don't want the ten sword-and-sorcery novels you bought for your nephew to influence Amazon's recommendations—but also gives users multiple paths from the book they know they want to the ones they don't yet know they want. Some paths are cleared by the brute ability of computers to pull together books into ad hoc categories undreamed of in Dewey's philosophy. When a publisher makes the full text of a book available to Amazon, it uses a set of algorithms to find uncommon phrases it calculates are important to the book. For *The Little House Cookbook,* the list of "Statistically Interesting Phrases" includes "sterilizing kettle," "pie paste," "pastry surface," "buttered pie pan," and "blood-warm water." Click on any of these phrases, and Amazon will show you other books that also use them: "Sterilizing kettle" turns out to occur in *The Fall: A Novel* by Simon Mawer. Amazon does a similar data-mining analysis on capitalized phrases that seem important to and distinctive of the book. Click on the link to "Laura Wilder" and you'll find a variety of books about her, including one on gender and culture in her writings. These statistically constructed paths through the geography of knowledge can take you to some unexpected territory: from "Laura Wilder" to *Complete Baseball Record Book, 2004 Edition* in just two clicks.

Listmania, one of Amazon's most popular ways of clustering books, relies entirely on manual effort. Hart explains, "You see features like the top ten movies of all time. Why not let people create their own?" Now there are hundreds of thousands of lists created by readers. On *The Little House Cookbook* page, the featured list at this writing is "Wisconsin Masters in the Arts," composed by Robert Schmitt, who identifies himself as "Former Wisconsinite." His list has twenty-three entries—no padding to get to a multiple of ten here—including books by and about Wilder, such as the *Laura Ingalls Wilder Historic Highway Guide: Wisconsin, Minnesota, Iowa, South Dakota, USA*. The list's page includes a sidebar of other lists Amazon has algorithmically decided are related, such as David Horiuchi's list of TV westerns, where you'll find the innocent *Little House on the Prairie* series alongside the scabrous *Deadwood*. Horiuchi's page lists yet more lists, including "Gifts for My Three Nieces" by Marty Shane, "proud aunt," who puts *Little House on the Prairie* in the company of "Barbie Primp and Polish Styling Head with Hands for Manicure," perhaps as far from *Deadwood* as one can get.

Amazon itself is about as far from a Dewey-compliant library as one can get. Dewey created a single way to cluster books; Amazon finds as many ways as it can. Melvil Dewey took the design of the system upon himself; Amazon lets anyone create her own category, give it a fun name, and publish it. Dewey prized neatness and order, bowing to the metric gods when he created a system based on multiples of ten; Amazon likes a friendly disorder, stuffing its pages with alternative ways of browsing and offbeat offers peculiar to each person's behavior. When you go to find a book in a Dewey-based library, you may be delighted to find another book on the same topic next to it on the shelf; when you go to buy a book at Amazon, the planned serendipity shows you a far wider range of books, determined by Amazon's editors, algorithms, and fellow shoppers. Dewey's system prizes the stability that comes with the physical world—books on bookshelves, white ink on spines; Amazon prides itself on its ability to cluster and recluster instantly.

These are differences not in the particularities of the categories and their arrangement but in the fundamental nature of organization. The Dewey Decimal Classification system is a second-order way of organizing, constrained by the physics of paper to give each book a single spot on a shelf, and having value because the system is as stable and reliable as the tonnage of paper distributed horizontally across library shelves. As a third order of order, Amazon is free of physics's onerous restrictions on the structuring and connecting of information. But the choice between being a Dewey or an Amazon isn't binary: Amazon displays its own Dewey-like categorization scheme at the bottom of each book's page as one more way that people might like to explore ideas.

Yet third-order organization is not a mere incremental improvement. In doing its business of selling books, Amazon overturns each of Melvil Dewey's three big ideas. First, while Dewey sought to find a single universal system to catalog books, Amazon provides a unique organization for each user. Second, Dewey arranged books by subject, but Amazon tries to find every way we might want to get from the A of a book we know to the B, C, and Z of books we don't know we're interested in, including the simple fact that lots of other people bought Z. Third, while Dewey liked the precision, predictability, and uniqueness of decimal numbers, Amazon throws books in front of your eyes with abandon. Compared to the neat row of numbered volumes on the shelf of a library, Amazon is a carnival of books, where even the orderly rows of the marching band are interrupted by a weaving conga line of suggestions.

But, then, Dewey's goal was to map knowledge. Amazon wants to sell us books. Its organization of its offerings is not bound by an underlying geography. Amazon is able to treat its enormous collection of books—that is, the books it can get if someone wants a copy—as a miscellaneous pile that can be digitally sorted to reflect the individual interests of each visitor. In the second order, the bigger a miscellaneous pile grows, the harder it gets to use. In the third order, piles offer exponentially more possibilities and more value the larger they get, as long as you keep them well and truly miscellaneous.

The fundamental problem with Dewey's system is not that he was an eccentric or that his early education was provincial. The real problem is that any map of knowledge assumes that knowledge has a geography, that it has a top-down view, that it has a shape. That assumption makes sense in the first and second orders of order. It unnecessarily inhibits the useful miscellaneousness of the third.

4

LUMPS AND SPLITS

It's a long drive. The kids are in the back seat. They've colored in their coloring books. They've listened to the CDs you brought for them. They've eaten their fruit snacks—the ones that contain 2 percent fruit and 30 percent sugar. You know they're getting edgy because they're starting to complain about each other. So you interject, in an overly delighted voice, "Let's play Twenty Questions!"

Although you just want to keep your kids quiet, by the time everybody in the car has had a chance to be "it," your children have learned some lessons.

They've learned what type of object is suitable to have people guess at: A desk, yes. Furniture, no. The brass knob on the guard's desk you happened to see in the Louvre in 1987, definitely not, unless you're actually trying to make your children cry.

They've learned how to hint, a sophisticated process that requires gauging not just how ideas interlock but what the hint will mean to someone who doesn't know all that you do.

They've learned the difference between hinting and cheating, and thus when it's okay to bend our own rules.

Perhaps most important, Twenty Questions has shown them that the world is organized so perfectly that we can get from ignorance to knowledge in just twenty steps. The game is called Twenty Questions and not Four Thousand Questions because—and this is perhaps the subtlest lesson it teaches our children—we've divided our world into major categories that contain smaller categories that contain still smaller ones, branching like a tree. That we can get from concepts as broad as animal, vegetable, and mineral to something as specific as a penguin's foot in just twenty guesses is testimony to the organizational power of trees.

In the third order of order, though, businesses such as Amazon increase customer satisfaction *and* sell more goods by willfully violating the perfect, treelike order of organizational systems, linking products across branches like a popcorn chain strung with abandon. As customers and businesspeople, we're learning lessons very different from those we absorbed playing Twenty Questions in the back seat of our parents' car. As was true then, the lessons go far beyond the game itself. They touch on the most basic ways we put the things of our world together.

THE SECRET LIFE OF LISTS

The better your children are at Twenty Questions, the more unprepared they're likely to be when they go off to college and encounter Jorge Luis Borges's essay "The Analytical Language of John Wilkins." There Borges invents a Chinese encyclopedia, the *Celestial Emporium of Benevolent Knowledge,* that divides animals into:

> (a) belonging to the Emperor, (b) embalmed, (c) tame, (d) sucking pigs, (e) sirens, (f) fabulous, (g) stray dogs, (h) included in the present classification, (i) frenzied, (j) innumerable, (k) drawn with a very fine camelhair brush, (l) et cetera, (m) having just broken the water pitcher, (n) that from a long way off look like flies.

A list is our most basic way of ordering ideas—the equivalent of lining up your shoes or hanging towels from a row of hooks. First one, then the other. If it got any simpler, it wouldn't be organized at all. By confounding our every expectation about lists, Borges shows us that there's more to them than we'd imagined.

Every list in our house, for example, has at least one thing in common. Whether it's a list of groceries to buy, things to do, checks we've written, friends' birthdays, credit card numbers, lock combinations, emergency telephone numbers, local movie theaters, plumbers we'll never use again, the heirlooms we think our siblings defrauded us of, or the names of the Seven Dwarfs, each of these lists is *about something.* We don't compile lists that mix types of spiders, constellations named after Greek gods, and the chronology of our dental work. Borges does in his list, but we don't.

Borges is able to ignore the "A list is a list *of something*" rule because he has violated an even more basic one: A list is compiled *for some reason.* We make a list of birthdays in order to remind ourselves to send cards and a list of restaurants that deliver so we can order food from them. Borges is in the unusual position of compiling a list to make a point about lists. The *for* of his list is "to be a list of things never found together on a list." It's as if your grocery list included celery, dish soap, and "timid" or "vertical"—they're not even the same part of speech as the rest of the list.

There's something else screwy about Borges's list. On our grocery list we do not have an entry such as "canned goods"—it's too broad a category. But Borges includes "stray dogs," "having just broken the water pitcher," and "innumerable" as if they were all at the same level of generality. We do have a rule for accommodating entries at different levels of abstraction, although not for entries as random as those on Borges's list. At our house, where we throw out ten pounds of newspaper every week but consider using a clean piece of paper for a shopping list to be environmentally unsound, we write our shopping list on the back of a used envelope. The list typically starts out as a list of groceries, but if someone decides we need a bottle of drain opener, she will write and underline the word "Hardware Store" and put "drain opener" beneath that, with the number of exclamation points indicating the urgency. If we need fresh crickets—we're vegetarians, but our pet frog doesn't adhere to our dietary principles—they will be listed beneath the underlined words "Pet Store." Those headings are information about the information that follows; that is, they're metadata.

There are rules for listing metadata, just as there are for listing items. The metadata should be differentiated in appearance, perhaps underlined and capitalized, perhaps written in a different color. The items have to be laid out to make clear their relationship to the headline, usually by writing them beneath their heading. This distinctive formatting reveals *nesting,* one of the most powerful ways of organizing ideas.

NESTS IN TREES

Historians don't agree about what is the earliest map. The one most often accorded the award was discovered in 1930 in Iraq and dates from 2500 to 2300 B.C.E., during the Babylonian dynasty of Sargon of Akkad. The palm-sized tablet shows a district of Ga-Sur, about three hundred miles north of Babylon, situated between two ranges of hills, with a river running through the middle. In the center is a plot of land owned by someone named Azala; the names of the owners of the other plots have become illegible. The hills are marked by overlapping circles, similar to the topographic markings we use today. The other contender is a nine-foot-long wall painting discovered in 1962 in Ankara, Turkey. It seems to depict eighty buildings in the city of Catal Hyük, where it was located, with a volcano erupting dramatically in the background.

There's no question that the Catal Hyük painting predates the Ga-Sur tablet by around four thousand years. But under any reasonable definition of "map," Catal Hyük is a marginal example at best, for the same reason that we don't count landscape paintings of the Hudson River as maps of New York. For something to be a map, we expect it to use symbols and have labels. Even better, it should show boundaries. The drawing of Catal Hyük does not.The map of Ga-Sur does and is indisputably a map. Indeed, it was probably drawn to demarcate those borders so that Azala's abutters wouldn't plant seeds in his field or the ruler could tax Azala. Even in our earliest clear example of a map we used nesting to understand our world: A farmer's plot of land is part of a hamlet, and a hamlet is part of a larger political unit, just as our towns are parts of counties, states, and a nation. If all politics is local, all localities are nested.

Nesting is a fundamental technique of human understanding. It may even be *the* fundamental technique, at least in its most primitive form: lumping and splitting. In the case of a map, a boundary *splits* off some unit of land and *lumps* together what's within the boundary. The same lumping and splitting that created a map of Azala's hamlet created Melvil Dewey's tree of knowledge. Indeed, Dewey's system could be translated into a map, with ten large continents, each divided into ten countries, and so on down to the smallest decimal number, represented perhaps as a particular room in a house at a particular address. Maps and trees are different representations of the same way of nesting information.

But to get from maps to trees took a genius. Before Aristotle we knew how to *use* nesting—language recognizes that robins and sparrows are types of birds—but we didn't know how to *think* about it. In Aristotle's *Metaphysics,* we can see him take the leap of thought required to understand how nesting works.

The *Metaphysics* is an inquiry into what it means for something to be. There is no simpler question. The title literally translated means "Beyond Physics," but it's not clear if Aristotle meant it to be taken as *What's Beyond the Physical* or *Physics: The Sequel*. Either way, in this work, often considered to be austere even for Aristotle, his patient connecting of fundamental concepts lends the work a quality similar to a Bach keyboard piece, beautiful in both its order and its unpredictability.

Already, in the fourth century B.C.E., Aristotle had a tradition of thought to contend with. He agreed with Plato, his teacher, that to be something is to be a particular type of thing: a robin, a man, a vase. But what makes a thing into that type of thing? As happens so often in life, Plato was misled by the examples he used. Plato turned to geometry, where students have long been taught that the triangle on their test paper is an imperfect representation of the "real" triangle the problem concerns. Plato extrapolated that to all things: The robins, men, and vases we encounter in life are but poor sketches of the perfect versions of each. These perfect versions are Plato's "forms," which he hypothesized we must encounter before we're born because we never see anything so perfect during our lives—a theory that has caused generations of freshmen to write off philosophy.

Plato's theory did not sit any better with Aristotle, a practical, reality-centered man who dissected over a hundred fish to see how they work. What is the actual relationship between a robin and the eternal form of a robin? Plato's word "participate" doesn't explain anything, Aristotle complains. "To say that the ideas are patterns that other things participate in is to use empty words and poetical metaphors," he writes. Originals don't "participate" in their copies. And even if we could explain how Plato's forms cause things to be what they are, Plato doesn't explain how a man can participate simultaneously in the form of humans, bipeds, and animals. Plato talked as if categories themselves were things. Imagine there are five hundred elephants in the world. For Plato, there is a 501st elephant: the category of elephant. In fact, in Plato's view the 501st elephant is even more real than the other five hundred because it is eternal and because the five hundred elephants become elephants only by "participating" in the 501st.

Aristotle understood, in a way Plato did not, that while categories such as "elephant" are real, they're not real in the same way that the five hundred elephants are. That took an epochal leap of understanding. Instead, Aristotle said that a category was a definition (or "principle") that explained why some things fit into it and others do not. If the definition of a bird is that it is an animal with two legs and feathers, penguins are birds but bats are not. Further, to be a bird means to be also in the animal category, which has its own definition. In addition, the bird category may have subcategories, such as water birds and jungle birds. This lumping and splitting continues until you get to individuals, the leaves of the tree. The result is a branching tree of categories in which each thing is simultaneously lumped with some and split from others. (It took another five hundred years to represent nesting graphically as a tree, an honor generally accorded to Porphyry, a Syrian-born philosopher in the third century A.D.)

With this principle of organization, Aristotle gave us a tool for understanding that "scales," as venture capitalists like to say: It works on the large scale as well as the small. It scales because it allows most of what we know about a thing to remain hidden. When someone tells me that an animal is a bird, I know without any further work that it is an animal, it has a backbone, it reproduces, it's mortal, it's a material thing, and more. I don't have to be told any of that. More important, I don't have to think about each of those things every time I see a bird, because I know the categories are available for my attention if needed. Trees are a supremely powerful way of understanding systems as complex as, say, the universe.

Aristotelian trees have persevered for millennia, and are present in the Dewey Decimal Classification system, the Bettmann Archive, Amazon's system of categories, the division of books into chapters and subheadings, the layout of health clinics into areas of increasing specialization, and the arrangement of items on restaurant menus. But trees come with assumptions embedded so deep in our tradition of thought that they look like common sense:

- A well-constructed tree gives each thing a place. If too many items don't have places and thus have to be shoved into the "miscellaneous" category, then the tree isn't doing its job.
- Each thing gets only one place. Listing the cheese plate under appetizers, entrées, *and* desserts just confuses people.
- No one category should be too big or too small. If your clothing catalog has a separate section for every shoe in every size, its organization will be too "bushy" to be of much use.
- It should be obvious what the defining principle of each category is. A list of real estate offerings that has a category called "Places" isn't very helpful.

When we're organizing a menu or a record collection, these are useful rules of thumb to make information more findable. But Aristotle began a long tradition of thinking that trees aren't just convenient, they're how the cosmos itself is organized. The tree of knowledge, the tree of species, the breakdown of the human body into major biological subsystems, the division of consciousness into reason and emotion, even the division of the earth into continents and countries—all are ways of *understanding,* not ways of looking up information. The passion with which we dispute the details of the trees we've constructed—Is Pluto a planet in the tree of heavenly bodies? Is homosexuality a syndrome in the tree of psychological diseases?—demonstrates that we believe, along with Aristotle, that some trees reflect the neat, clean, comprehensive, knowable branching structure of reality itself.

All along, though, our knowledge of the world has assumed the shape of a tree because that knowledge has been shackled to the physical. Now that the digitizing of information is allowing us to go beyond the physical in ways Aristotle could not have dreamed, the shape of our knowledge is changing.

LAUNDRY AND LINNAEUS

Aristotle lays out a task for all those who want to know their universe: Go forth and lump and split.

"Lump" and "split" are not Aristotle's words, but, surprisingly, they are technical terms among professional indexers. Seth Maislin, a member of the board of directors of the American Society of Indexers and a consultant on indexing to the likes of the United Nations and Microsoft, explains: "A lumper takes things that seem disparate and combines them because they have something similar. A splitter tends to take two things that are lumped together and separate them into smaller categories." Indexers tend to be one or the other, their technique driven by their personality.

Every day we face the same choices as professional indexers. Some of us store all our bed linens in one pile in the closet, while others of us separate them by bedroom, color, weight, and season . . . and then arrange each little pile so the least-worn sheets are on top. When asked if we know the way to San Jose, our directions lump together a long stretch of road rather than counting the precise number of lights—"just keep going for a while"—but split out the strip mall on the right because we think it will be a useful landmark. And much of our conversation is about the right lumping and splitting. Your friend didn't like the movie last night because she thought it was supposed to be a comedy. No, you say, it wasn't supposed to be a laugh-out-loud comedy. It was more of a chortle-inwardly comedy . . . more like *Amélie* than *Animal House*. We are constantly negotiating life's lumps and splits, from trying to decide which kid gets to ride in the front to arguing over health-care reform.

The remarkable fact is that we have built systems for understanding the universe using the same technique we use for putting away our laundry: Split the lump of cleaned clothes by family member, split each family member's lumps by body part, then perhaps split by work or play, by season, or by color. If you're not sure whether the ski socks go with the normal socks, the winter wear, or the sports wear, you're still going to have to pick one because they have to go somewhere, and they can go only one place. That's just how atoms work. At the end of the process, you will have created a conceptual tree of clothes, with each family member as a main branch, the body parts as subbranches, and the ski socks hanging from whichever branch you finally chose. That's how we get through laundry day. Yet Linnaeus, the father of our modern way of organizing nature—the person who split the universe into the animal, vegetable, and mineral lumps that start almost every round of Twenty Questions—used those very same laundry-sorting principles to describe the structure of the entire natural world, from rocks to primates.

Born in 1707 to a pastor and his wife in the small town of Stenbrohult, Sweden, the young Carolus Linnaeus was fascinated with botany. He became a physician, but he spent most of his life devising and applying a system for classifying natural objects. His way forward was cleared by a heroic act of lumping accomplished by the French naturalist Joseph Pitton de Tournefort thirty-five years earlier. Tournefort introduced the notion of the *genus* (plural: *genera*), clustering the six thousand known plant species into just six hundred groups. That pared down the number enough so that Linnaeus could classify plants by looking at the shape, number, relative size, and arrangement of their stamens and pistils, the plants' reproductive organs. Linnaeus's system had enough variations—5,776 by his calculations—to let it account for all the plant genera. Because the system utilized easily observable characteristics, with just a little training anyone could examine a plant's parts and know where it fit. And because the parts of a plant come in identifiable and countable units, there were no messy borderline cases.

Linnaeus's system worked. But, even though he was the son of a clergyman, and even though he knew Genesis by heart, Linnaeus did not believe that the system of classification he had published revealed God's order. Although at times he hinted that it did—nature, after all, was assumed to be a book written by God—Linnaeus seemed comfortable with the idea that he had spent his life devising an order that was useful if not true. This was a century before Darwin showed how animals could be grouped not by mere similarity but by causation: Humans and chimps go on the same branch because the chain of causality leads back to a common ancestor.

It's hard to imagine today what it meant to categorize living creatures without evolutionary theory. With evolution, the tree of species is a family tree. But if creatures didn't evolve one from another, then putting them on the same branch signified only that they *resembled* one another. Is a bat a type of bird, because it flies, or is it a type of mammal, because it has fur? Linnaeus didn't think he could resolve such questions finally because he could not read God's mind. But, like Melvil Dewey, Linnaeus believed in the efficiency of rationality. Instead of advocating spelling simplification, Linnaeus came up with a highly efficient and orderly way of naming and organizing species, so scientists could agree on what species they were talking about, a condition for scientific progress.

Linnaeus also promoted the "binomial" system of naming, replacing names such as *Grossularia, multiplici acino: seu non spinosa hortensis rubra, seu Ribes officinarium* (the European red currant) with two-word phrases such as *Ribes rubrum.* The first word of a binomial is the genus and the second indicates something specific to that species. So humans are *Homo sapiens* ("men who are wise") and the painted sage plant is *Salvia viridis* ("a healing plant that is green"). Aristotle himself laid the groundwork for binomials in the *Metaphysics* by pointing out that if a classification in the tree is right, anything more than a genus-species name is redundant: There's no need to call Socrates a "human biped animal," because all bipeds are animals. (Carolus Linnaeus's own name—his father changed the family name from Ingemarsson to the Latin Linnaeus in honor of a three-trunked linden tree in their yard—is itself sort of a binomial, albeit in species-genus order.)

But a system so important in the history of natural science can't be explained purely through whims of personality. Linnaeus had a political aim: He—like Dewey—wanted to democratize knowledge. Linnaeus was a doctor and a teacher committed to spreading knowledge far and wide. Twice a week in the summers of the 1740s, he would lead up to three hundred people, including women, on twelve-hour natural history walks. Linnaeus's method of classifying plants was easy to teach and didn't require special equipment. To determine a plant's class, you first check the stamens and pistils. If they're in the same flower, it's "monoclinous," which translates to a marriage in which the husband and wife share a bed. If there is one stamen and one pistil, it's a monander—one husband in the marriage. If there are two stamens, it's a diander—two husbands in one marriage, and so on. Even Linnaeus, the pastor's son, knew: Sex sells. In fact, one German botanist declared the system's reliance on sexuality immoral.

You can see one more inspiration for his system by visiting the headquarters of the Linnean Society, in London. The society's entrance is tucked away in a courtyard shared with the far larger Royal Academy of Arts. A set of statues outside the Royal Academy provides a small tree of intellectual heroes: Cuvier and Leibniz to Linnaeus's right, and Newton, Bentham, Milton, and Harvey above him. Inside, the headquarters are very British nineteenth century, done in mustard and parchment, wood and brass. The Swedish patriot's lifework is here because Linnaeus's widow sold it to a rich young British scientist eager to make his mark; she needed the money for their daughter's dowry.

On the first floor of the building's library, heavy cloth covers a couple of glass-topped tables that exhibit some of Linnaeus's original specimens, the husks of the beings Linnaeus held in his hand when he said, "I name thee . . . thus!" They serve as the reference points for disputes about whether a particular binomial refers to this or that creature. The official brochure notes that the collection includes 14,000 plants, 158 fish, 564 shells, and 3,198 insects. The collection room itself is belowground, protected by a six-inch-thick metal door and designed to survive a nuclear bomb. "The whole of the taxonomic world depends on the legal concept of the *type*," Gina Douglas, the society's librarian and archivist, explains. It makes sense to bury first- and second-order organizations such as this one and the Bettmann Archive. Specimens made of atoms are fragile and need protection.

Inside the vault, the room feels oddly homey for a bomb shelter. It's only about fifteen feet square, but the wood and brass of the specimen drawers and bookshelves that line its walls lend the cramped space the air of a reading room at a gentleman's club. Douglas spreads out some framed specimen pages, each with one plant specimen, gray as dust, attached. "Notice the *K* on that one," she says, pointing to a small letter at the bottom of the page. "That tells us who collected it. It's rare for a page to have that information." Too bad, because some of the specimens in the cases upstairs had been misidentified. The note card for *Solanum quercifolium* explains that Linnaeus grew the plant from seeds he thought were from Peru but were actually from somewhere near Mongolia. If he'd had the name of the collector—important metadata—he might have avoided that mistake.

Douglas opens a first edition of *Systema Naturae,* in which Linnaeus dared to classify all of nature in just eleven pages. Of course, Linnaeus had to make the book the size of a small coffee table to fit all of nature into it. Douglas gingerly turns the pages—like turning down the sheets on a bed—to reveal the three double-page spreads at the tree's root: animals, vegetables, minerals. Linnaeus has caged the animal kingdom in six major boxes, four for vertebrates (mammals, birds, reptiles, and fish) and two for invertebrates (insects and a category called "worms" that included everything else—a squirming mass of the miscellaneous). In the upper-left box for animals with backbones is a box for mammals, topped by primates, into which Linnaeus—radically—put monkeys next to humans, referring to orangutans as *Homo sylvestris*, "feral man of the woods."

Although the boxes are nested, Linnaeus maintained his version of the Great Chain of Being, ranking each of the species within them. The method is like having separate classrooms for the advanced, average, and slow students and arranging the chairs in each according to the individual student's grade point average—or like Melvil Dewey clustering books by topic and subtopic, but still assigning each a decimal number so they can be laid out in neat rows on shelves. The principle of Linnaeus's ordering of the species is harder to figure than a grade point average, however. The medieval version of the Great Chain sorted creatures by how much "spirit" they had versus how much matter, putting angels above humans, humans above oysters, and oysters above rocks, which seems intuitively right. Linnaeus used a more worldly criterion—complexity—to come up with an ordering that gets the angels-humans-oysters-rocks ordering right. But complexity is itself a complex notion. Are we sure that rats are more complex than peacocks and that caterpillars are more complex than willows?

A hidden hand guided Linnaeus, determining the general shape of his scheme. Douglas withdraws a thin pile of paper cards as soft as handkerchiefs from one of the drawers. On each, Linnaeus has recorded in his fine hand the name of one species. If you have one species per card, you do what we've seen Mendeleev did with the elements: You play solitaire. You lump and split the cards, putting them near other cards like them. As you do so, you are drawing yet another map of knowledge. Linnaeus named the largest units in his classification kingdoms not because animals, vegetables, and minerals lord it over the creatures within their borders but because kingdoms are the most inclusive territories on political maps; Linnaeus explicitly likened the five levels of his classification system to "Kingdom, province, territory, parish, village" and hoped to see the three kingdoms "depicted in maps or paintings, printed under the title *Geographica Naturae*." That's why *Systema Naturae* is oversized: a map makes the most sense when you can see it all at once.

Linnaeus's system not accidentally shares properties with the paper that expresses it: bounded, unchanging, the same for all readers, two-dimensional, and thus only with difficulty able to represent exceptions and complex overlaps, making all visible in a glance, with no dark corners. Linnaeus's organization took the shape it did in part because he constructed it out of paper. Indeed, he classified plants by their stamens and pistils instead of by the more obvious sign, their flowers, in part because he wanted to be able to publish illustrations in black and white. That way the cost of the books would be low enough for the multitudes. Linnaeus wasn't putting away clothing, but because he used paper—atoms—to think through the order of the natural world, the organization he came up with repeated the general shape of an orderly household on laundry day.

TREES WITHOUT PAPER

So, what would a nested order look like if we didn't have to write it down on paper?

Pat Howard, vice president of strategy, marketing, and operations for IBM Business Consulting Services in the Americas, had a headache. He needed to put together highly qualified teams for projects that frequently spanned multiple countries. With 25,000 consultants to draw on, he might have found the task manageable if there were only one or two criteria to look at. But there were dozens. "If the lowest-cost qualified individual isn't available today but is available three weeks from now, am I willing to pay a little bit more for someone with the same qualifications or do I need to defer the start date?" he asks. "Or if I find somebody who's in the right geographic area, does the reduction in cost of travel offset the savings from using a lower-cost country?" How about language skills? Expertise in particular products? Certification in this or that banking system? And if he can't find one person who speaks French fluently and knows the SAP software package, can he build a team that hits all the project's requirements?

That takes more than a typical text search engine. Search engines can only tell if a particular string of letters is used on a particular page. They can't tell if a date signifies when a consultant is busy or when she's free, and they will miss all the French speakers in France who thought it was just too obvious to state explicitly that they speak French. Instead, IBM could let managers find consultants by clicking through some type of tree. Perhaps at the first level it would look like a map of the world. Click on a continent and you are shown a list of application areas in which consultants are qualified. Among those areas choose, say, accounting and you see five price ranges for consultants' day rates. Choose a price range and see the weeks when consultants in that range are available. As you climb out further on the branches of this tree, you're getting closer and closer to finding the perfect member of the perfect team you're putting together . . . unless you don't need to choose a consultant by continent because your client doesn't care about travel expenses and wants the best accounting consultant regardless of where she lives. In fact, your client has told you to get the top people and hang the expense! So you'd rather have the first branches of the selection tree show you areas of expertise instead of location. You don't ever want to see the price ranges, but you desperately want to sort consultants by their years of experience. Unfortunately, the existing tree doesn't get to that until step twelve.

You're stuck. What you really want is a tree that arranges itself according to your way of thinking, letting you sort first by area of expertise and then by experience, and then tomorrow lets you just as easily sort first by language and then by cost, location, and expertise. You want a *faceted classification* system that dynamically constructs a browsable, branching tree that exactly meets your immediate needs. That's precisely how IBM's consultant database works. According to

Howard, the new system enabled them to put together a project team for a very large insurance company in two days instead of the usual several weeks. Benefits like that explain why the *New York Times*, Barnes and Noble, and NASA are implementing faceted systems. Endeca, one provider of faceted classification systems, has a customer in the oil and gas industry that uses its software to provide access to 25 million different items sorted on more than a thousand different properties. Siderean, a competitor, is working on systems for the financial services and health insurance industries that will provide faceted access to up to 200 million customer service records and transactions.

Faceted classification combines the user-friendliness of browsing a tree with the power of digital computing. It is unthinkable without computers. So it's surprising that it was invented seventy years ago by a librarian inspired by a mechanical toy, decades before the age of the computer.

S. R. Ranganathan was born a Brahmin in the tiny town of Ubhayavedantapuram, in southern India, in 1892. By the time he started college, at seventeen, he had been married for two years to a girl who at the time of their wedding was all of eleven. He became a teacher of mathematics and physics with so little interest in librarianship that he applied to become the University of Madras's first librarian only because the money was better—and even then his friends had to urge him on. But after a trip to London to study library science—a term he coined—libraries became his life. They seemed to appeal to his orderly mind and his spiritual desire to help people. Ranganathan's "five laws of library science," published in 1931, gave voice to both:

> Books are for use.
> Every reader his/her books.
> Every book its readers.
> Save the time of the reader; save the time of the library staff.
> The library is a growing organism.

The Dewey Decimal Classification system was widely used in India at the time. Ranganathan wanted a new system with a point of view more relevant to India than Dewey's Christian worldview and with more flexibility built into it.

He was in London when he had his breakthrough. "I happened to see a Meccano set being demonstrated at a Selfridges store," he later wrote. (A Meccano set, like an Erector set, lets youngsters build machines.) "I spent a whole hour observing how different types of toys could be assembled from a small set of basic components." He proposed five basic areas of categorization, or facets: personality, matter, energy, space, and time. For each he came up with a list of possible values, which he called "isolates." By combining the isolates for each of those five facets, books could be flexibly classified without having to spell out ahead of time every possible slot. It is not a simple method: A book on the management of Indian banks up to 1950 would be expressed as "X62:8.44'N5": X for economics (personality), 62 for banks (matter), 8 for management (energy), 44 for India (space), N5 for 1950 (time). In 1933, he published his masterwork describing the system, *Colon Classification*. Yes, it's the world's worst title, but the system it outlines for classifying books—using categories separated by colons—was revolutionary. It immediately sold out its initial printing.

Colon Classification reflected Ranganathan's personality. He was a meticulous man who lined up all the ingredients before starting to cook. When preparing for a trip, he weighed each item before packing it to keep under the limit. His daily schedule was so predictable that a burglar once scheduled a successful fifteen-minute incursion. Like Melvil Dewey, he was preternaturally irked by avoidable inefficiencies, such as waiting for books to be fetched by librarians. Yet Ranganathan also had a strong nonrationalist side; he dabbled with Ouija boards and visited swamis, including one, Swami Swayamprakaasa, who lived in a cave and, according to Ranganathan's son, "wore nothing on himself except a long beard and a *rudraaksha,*" which, he helpfully explains, is "a garland made out of some marble-sized nuts." Ranganathan expected librarians to have a spiritual bent, using their intuition to categorize books. With intuition, he wrote, a person "sees beyond the phenomenal occurrences. He transcends space and time. He sees from the seminal level, the perfect harmony of everything." Intuition introduces vagueness into his system; according to his son, "Even Ranganathan was apparently not very clear about what the 'Personality' facet really stood for."

Yet Ranganathan's mysticism-tinged system has a property that makes it remarkably powerful when used in computer-based systems: No facets have to be assumed to be the "root." Rather than deciding ahead of time what the "proper" tree is, the computer can construct a tree on the fly based on the user's interaction, just as team managers do with the IBM consultant database. Start with the facet most relevant to your current interest, and then limit it further by using another facet, until you find what you want. You get to play Connect the Colons any way that suits your needs.

Of course, personality, matter, energy, space, and time—Ranganathan's original facets—don't make much sense if you're organizing parts for the oil and gas industry. What exactly constitutes the "personality" of a three-millimeter grommet? Rather, each collection has its own facets and appropriate values; often the facets and values can be pulled automatically from an existing database. The result is a system that lets us become data squirrels, jumping from branch to branch . . . and wherever we jump, a branch magically appears.

This is not a trivial computing problem. Steve Papa, CEO and cofounder of Endeca, one of the leading companies providing faceted classification systems, takes visitors to his Boston office through a demonstration the company built using ninety thousand reviews from *Wine Spectator* magazine. Each review can be sorted on any of nine facets: wine type, country, winery, rating, price range, year, special designations, drinkability, and flavors. Since you can sort these in any order you want, there are 10^{34} logically possible paths—but users see only paths that end in existing types of wine. For example, if you ask to see highly rated wines, the wines for under five dollars a bottle go away because, unfortunately, there's nothing on the great-but-dirt-cheap branch of the tree. Likewise, ask to see the Zinfandels and all the countries except the United States and South Africa disappear. In a faceted system, there are *no dead ends* down which we may accidentally wander. That's why engineers trying to find just the right part in a database of 25 million possibilities depend on faceted classification.

Endeca also developed a product that seems far removed from the faceted classification system: data reporting and visualization. For example, the company is working with Harvard University to roll out a system that enables a thousand people in alumni relations to generate reports about donations, using twenty different facets. Papa demonstrates by creating a chart from three facets that show how much was donated in a particular year, broken down by the age ranges of the donors. When he selects a different facet, the report updates to show the regions where a fund-raising lunch is likely to be most lucrative. This is exactly the same faceted information that Endeca uses to construct browsable trees on the fly. Faceted classification can be used either way because it captures something important about the organization of the real world that organizational trees do not: Reality is multifaceted. There are lots of ways to slice it. *How* we choose to slice it up depends on *why* we're slicing it up.

Over the course of time we have largely let go of our Aristotelian belief that there is only one right and true tree of knowledge, but we have behaved as if the rule was still in effect because we have had to use atoms—usually paper—to preserve and transmit information. When you organize knowledge by arranging slips of paper, you get trees that have one place for each leaf. When you draw the shape of knowledge on pages, you draw neat borders and don't have room for ambiguity. When you publish knowledge in books, you put those ideas into a treelike structure of volumes, books, chapters, sections, paragraphs, and sentences. Implicitly, paper shapes knowledge into trees.

Now that we have a paper-free order of order, we're certainly not going to give up nesting categories. Without nesting, language wouldn't have gotten past pointing and grunting. Try defining "jogging" without referring to it as a type of running, or "martini" without having to admit that it's a type of cocktail. Nesting is here to stay. But trees are a different issue. Of course there are many times when the path through knowledge is shaped like a tree. But not always. Trees don't work that well when a history of military cooking should hang from the history, military, *and* cooking branches. Lumping and splitting physical objects requires us to make binary decisions about where things go. Ideas, information, and knowledge shouldn't have to suffer from that limitation.

In the third order of order, a leaf can hang on many branches, it can hang on different branches for different people, and it can change branches for the same person if she decides to look at the subject differently. It's not that our knowledge of the world is taking some shape other than a tree or becoming some impossible-to-envision four-dimensional tree. In the third order of order, knowledge doesn't have *a* shape. There are just too many useful, powerful, and beautiful ways to make sense of our world.

5

THE LAWS OF THE JUNGLE

I'm ashamed to say I sometimes empty the dishwasher of all its contents except for those in the silverware basket, hoping that my wife will do it. It's a miracle our marriage has lasted so long.

I don't mind putting away the plates because I have a technique. I swoop in from the top and take as many of the dinner plates as possible, then pick out the salad plates, and continue on, from largest to smallest. That's not because I'm an obsessive-compulsive who has to put things away by size order. If anything, it's reality that's obsessive-compulsive. The bigger plates insist on sticking up higher than the smaller plates, and thus are easier to take out first. The silverware, on the other hand, refuses to play along. It mobs up in the dishwasher's basket, unsorted and defiant. I grab fistfuls and have to go through them item by item. The silverware basket is too damn miscellaneous.

Of course, it's only a problem because we insist on "rescuing" the silverware from its miscellaneousness, dividing it among the designated sectors of the silverware drawer. If we could leave it as miscellaneous, we'd just dump it into the drawer and be done with it. That, in fact, was our Silverware Maintenance Policy when I was living with three other college students. But we paid for our cavalier refusal to split the silverware lump every time we went to set the table. We would paw through the drawer, picking out pieces we needed. If we're going to use utensils to eat—a nicety we sometimes skipped in my college days—then one way or another, we have to take the silverware out of its state of pure miscellany. In college, we postponed that moment until we were ready to eat. As responsible married adults, we now sort the silverware in advance of its use so that when we need to set the table, the silverware is tucked away in its little beds, waiting for us.

Each strategy has its place in the world of atoms. The college-student approach works best when you have only a few types of silverware. Once you've gotten flatware as a wedding gift and you begin to add to your collection, you find you don't just have knives, forks, teaspoons, and soup spoons. You have butter knives, steak knives, paring knives, and service knives. You have salad forks, regular forks, and forks for eating shrimp. Then you get older and have silverware for the family, for guests, and for super-special guests. And if you're lucky enough, like some of us, to marry an Orthodox Jew, you have separate sets for meals with meat and meals with dairy, and double that for Passover. At that point, making one giant heap of all your silverware while cranking up Jefferson Airplane on the hi-fi no longer seems like such a good idea. If you don't sort after each meal, your silverware world falls into chaos.

In the digital age, computers have become demonically good at sorting through gigantic, complex piles of information. Crate and Barrel's online catalog has fifty types of place settings, each with two different forks, two different spoons, and a knife, yet there you can find a pierced serving spoon that goes with the genuine-pakka-wood-handled flatware faster than you can grab a bunch of soup spoons from your perfectly ordered silverware drawer. And that means we college students had it right. We were just ahead of our time. The best digital strategy is to dump everything into one large miscellaneous pile and leave it to the machines to find exactly the table settings we need for tonight's dinner.

A BIG CAN OF WORMS

"Alison Lukes et Cie is Washington's premiere closet consultant, personal shopper and stylist." Her Web site shows her sitting in front of a closet in which dozens of women's shoes on shelf after shelf are arranged by color. She is young. She is stylishly dressed. She is pretty. She is smiling proudly. And why not? Ms. Lukes et Cie wrestle one of the last remaining pockets of household miscellany into order. Closets crammed with junk shrink back in terror when they hear her dainty tread. Chaos itself quails before her label gun.

What is our problem with the miscellaneous, anyway?

At its heart, the miscellaneous is a set of things that have nothing in common. Of course, that "nothing" is relative since the utensils in your kitchen's miscellaneous drawer all have a use in preparing and eating food, all are physical objects, and all are smaller than the drawer itself. Likewise, the miscellaneous section of a greeting-card display does not hold bassoons or riding lawn mowers. Nevertheless, within some particular domain, the miscellaneous gathers things that are unlike whatever sits next to them.

Sometimes we like that. Sometimes, as with college housemates, sorting on the way in takes more effort than sorting on the way out. And there can be positive benefits to miscellany. Enlightened human resources directors will tell you that workplace diversity isn't just a matter of equity. Pepsico says that about an eighth of the company's revenue growth in 2004 came from new products "inspired by diversity efforts." Putting unlike things together also works for Oscar-winning film editor Walter Murch. When he was editing *Jarhead,* he filled the walls of his studio with jumbled photos of the five thousand separate shots in the movie. "It makes images collide with each other in very opportune ways," he said.

Nevertheless, when Murch is done with a project and is filing away the photos, he undoubtedly wants to sort and order them. In the first and second orders, mixing things up may be great for creativity, but for refinding them, it's a disaster. As one of the "Slob Sisters" writes in their popular book *Sidetracked Home Executives:* "On that fateful June day, I was in my new home, lying on the living-room floor, surrounded by 157 Belkins' moving boxes—all marked MISCELLANEOUS." That's not a good thing.

Likewise, in classification systems, an overstuffed miscellaneous category can be a sign that the system isn't using all the relevant information. If I cluster elephants with spotted owls, the basis of the clustering—the category's Aristotelian definition—adds the information that both are endangered species. But if I throw them both into the miscellaneous category along with dachshunds and crickets, I've buried that relationship. So we should be suspicious (as Stephen Jay Gould brilliantly pointed out) when a taxonomic system divides a domain into two major lumps that are wildly uneven—as Linnaeus's classification of animals into vertebrates and invertebrates did. He divided the vertebrates (which we now know includes forty thousand species) into four subcategories, and the invertebrates (about a million species) into just two: the rather well-defined category of insects and the undifferentiated mass of creepy-crawly-swimmies—everything from earthworms to clams and jellyfish—which he called "worms" (*Vermes* in Latin). Linnaeus didn't know that his system was so skewed—he thought there were fewer than fifteen thousand species in total—but he nevertheless paid inordinate attention to the animals that, like him, have backbones.

It isn't obvious why Linnaeus pushed so many species into the *Vermes* bucket. While it's true that the simpler the organisms, the fewer ways to differentiate them, Linnaeus broke plants down in intricate detail, even though they are less complex than worms. Perhaps it's simply that Linnaeus loved plants and just didn't care much for worms. Whatever the reason, he left too much information hidden in that miscellaneous bucket.

It was Jean-Baptiste Lamarck—unjustly remembered primarily for being wrong about how giraffes got long necks—who not only sorted out Linnaeus's worms but changed the basic shape of Linnaeus's tree.

As Gould recounts the story, Lamarck loved invertebrates so much that when he was almost fifty, he was appointed professor of insects and worms at the Muséum National d'Histoire Naturelle. Lamarck called *Vermes* "a kind of chaos where very disparate objects have been united together" and, starting in 1793, he began to draw distinctions. When, in 1802, he split the Annelida—earthworms and the like—from intestinal and other host-based worms (Linnaeus's decision not to delve deeply into this category will strike many of us as increasingly understandable), *Vermes* was left with creatures not obviously more complex than the category of sea urchins supposedly below it. Rooting around in the bucket, Gould claims, Lamarck came to realize that life could not be ordered in a single line, from least complex to most, as Linnaeus had.

There are two different lessons we could draw from Lamarck's correction of Linnaeus's system. We might say that the miscellaneous category should make us wary because it hides information waiting for a Lamarck to come along and split the lump in useful ways. Or we might say that Linnaeus was not miscellaneous enough. Sure, Lamarck discovered important distinctions among worms. But every time you organize matters in one way, you are disordering them in others. Sorting my dessert recipes into cakes, cookies, and pies obscures their carbohydrate order.

The basic fact that order often hides more than it reveals has sometimes itself been hidden within the art and science of organizing our world. We have been like the proverbial seven blind men feeling the elephant, except unlike the narrator of the story, we've had to pick our favorite blind man. Lamarck's division of *Vermes* indeed reveals relationships Linnaeus missed, but a fisherman would divide the bucket still differently based on which wriggling creatures desirable fish are likely to snap at. Lamarck's and the fisherman's divisions both have merits, but if it's a first-order bucket, we can divide it only one way, just as Staples has to shelve its printer inks one way and not another. In the second order, we have the flexibility to organize physical metadata in a few ways—library catalog cards sorted by subject, author, and title—but not much beyond that or the catalog gets too big to be usable.

These physical limitations on how we have organized information have not only limited our vision, they have also given the people who control the organization of information more power than those who create the information. Editors are more powerful than reporters, and communication syndicates are more powerful than editors because they get to decide what to bring to the surface and what to ignore.

At least in the first and second orders of order. In the third order, bits rule. And so does the miscellaneous.

TAGGING LEAVES

There's something glorious about a well-crafted, treelike structure of information, even if it sometimes borders on the absurd.

The list of categories in the International Press Telecommunications Council list of "NewsCodes"—a Dewey-like system for categorizing news articles—in spots accidentally reads like headlines that encapsulate a story: "Financing and stock offering." "Government contract." "Global expansion." "Insider trading." Such accidents of meaning happen, especially when you're trying to cover every topic about which a newspaper might want to write. The list's real peculiarity is how uneven its specificity is. While under "Cinema" there is a single entry ("film festival"), under "Sailing" there are seven sorts of dinghy races, including "one man dinghy (4.57m/sq mainsail)."

The Getty Art and Architecture Thesaurus is even more ambitious than NewsCodes. "The aim was to classify the material world," says Joseph Busch, the project director who oversaw its construction. "Quite a modest project," he adds, playfully. The thesaurus was created to enable institutions to find and share information about their contents, a boon for curators and scholars. The experts who accomplished this wonder of the well-organized world divided 128,000 terms into seven top-level categories. Search for "apple corer" and you'll find that it's part of the Corers category, which, through several shoots and twigs, eventually attaches to a main branch of the tree:

corers
culinary tools for extracting
culinary equipment for preparing and cooking food
culinary equipment
equipment by context
equipment
tools and equipment
furnishings and equipment
objects

Because the designers wanted to provide ways to describe artworks that depict scenes in which things happen, the thesaurus includes an Activities facet at the same level as the Object facet. Under Activities, you'll find the subcategories Standing, Sitting, Whispering, Archery, and Torture, all on the same level and hanging from the same branch. Classifications make strange bedfellows.

The Getty thesaurus provides what's called a *controlled vocabulary,* so curators can classify their holdings without having to make arbitrary decisions about whether to describe a maritime painting by J. M. W. Turner as depicting ships or boats. Standardization makes it easier to retrieve information: If you know the vocabulary (or if you browse the tree) you don't have to guess whether to use the word *ship* or *boat* when looking for that Turner painting. The fact that the Getty's terms are arranged into trees also helps avoid ambiguity, because the system knows that the word *rock* is a stone in the Objects branch, but *rock* is what Whistler's mother does in the Activities branch.

The Getty thesaurus is a mighty tree but, like all such projects, it can strive for comprehensiveness only by reducing the richness of what it's comprehending. This is the nature of organizational trees, for they are built on single relationships applied over and over again: "B is a type of A," or perhaps "B reports to A" or "B is the child of A." No matter the relationship explained by the branch, it is almost certainly too simple to capture all of the relationships and complexities of its subject.

Usually we know that. We understand that a genealogical tree expresses only the path of DNA through time and that it tells us nothing about the emotional ties among the children and parents. We understand that the Getty thesaurus is intended as a convenience for curators, not as a comprehensive guide to how the universe works. Even so, when we draw a map of knowledge, it is all too easy to assume that knowledge is a territory that can be subjugated by applying a rigorous and relentless methodology.

The Getty project was only practical because it was undertaken by a single organization that could make the hard decisions. This explains the problem with ladies' pants. Browse online at the Gap and you'll find pants, jeans, and capris. Capris are split into cropped pants and cropped jeans, with some messy overlap between the three sets. J.Crew has pants, loungewear, denim, and suiting, and splits the pants and denim groups into different leg cuts—matchstick, hip-slung, bootcut, boy jeans, slouch, city, favorites. Anthropologie has pants divided into wide leg, slim leg, trousers, denim, short pants, petite, and tall. Even browsing all the pants doesn't necessarily get you all pants: The Gap seems to put capris in the gap between pants and shorts. It doesn't get any better with skirts: Anthropologie does graphic, short, straight, fluffy, and petite skirts, whereas the others have short, long, and suit. And all the stores have sale sections, of course. If there were a controlled vocabulary and a standard tree of pants, shoppers could know how to browse every store, confident they're not missing the capris. It hasn't happened not because ladies' pants are more complex than the set of everything that might show up in an artwork but because there is no single entity with the Getty's standing to declare a standard classification scheme. Classification is a power struggle—it is *political*—because the first two orders of order require that there be a winner.

The third order takes the territory subjugated by classification and liberates it. Instead of forcing it into categories, it *tags* it. Tagging lets a user of online resources—Web pages, photos—add a word or two to them so she can find them again later. The basic idea has been around for decades, but one particular site, Delicious.com (also spelled "del.icio.us"), gave it a twist that sparked a new round of interest. Joshua Schachter, Delicious's creator, calls it an "amplification system for your memory of Web sites." On its most basic level, Delicious is a bookmarking site that lets you list Web pages you may want to go back to, especially if your list no longer fits comfortably in your browser's bookmark menu. To help you find the sites you've bookmarked, Delicious lets you attach whatever words you want to them. If it's a page about San Francisco, you might tag it "San Francisco," "SF," "my hometown," or "4-syllable cities." Tags let you remember things *your* way. When you want to refind sites that talk about San Francisco, you just click on the list of tags displayed on your own page at Delicious.com and it shows you a list of all the sites you've given that tag.

Tagging grew out of a very personal need. Schachter's own list of Web addresses had grown to twenty thousand, many of which he wanted to share with friends. So he built a site, now defunct, called Muxway, where friends could see the sites he'd listed. But because his friends were finding interesting sites, too, Schachter opened up Muxway so others could contribute sites they'd found. In 2003, Schachter, who was working as a financial analyst during the day, took what he'd learned from Muxway and built Delicious. Until he sold it to Yahoo! a few years later, he ran it from his apartment.

Tagging was the most important feature Schachter added to Delicious. The idea goes back to his original list of twenty thousand Web addresses. Just eight lines into the list, Schachter had annotated a site's Web address with "#math." By using a hash mark (#) to flag tags, Schachter could easily search the list specifically for them. At Delicious, of course, users don't have to type a # when they want to create a tag; when you put a bookmark onto your Delicious page, a simple form pops up.

Instead of using tags, Schachter could have set up Delicious so that users create folders into which they drag Web addresses, much like the typical Internet browser's bookmarks and like our computer desktops. But folders have a big disadvantage over tags: An item can go in only one folder, just as a physical book can go on only one shelf of a library. True, advanced computer users know that they can create what Windows calls a "shortcut," which allows you to put links to a file into multiple folders, but it's a time-consuming process that can quickly clutter a desktop. If you want to file a page about Aruba under "Aruba," "Caribbean," "beach," "vacation," "snorkeling," "trips," "too expensive," and "daydreams," you'd have to make a folder for each term. At Delicious, you'd simply type in those terms when you bookmark the page. Each of those tags then shows up in the tag list on your Delicious page, and clicking on any one of them assembles a list of all the pages you've tagged with that word. You can also find all the Web sites you've tagged with both "beach" and "vacation," which would exclude the pages about Greenland you tagged as "vacation" and "extreme."

Think how different this is from the Getty thesaurus. Rather than using a standard set of categories defined by experts, at Delicious—and the many sites that followed its lead—each person creates her own categories in the form of tags. The Getty's categories are carefully nested, creating a well-ordered tree that would have made Dewey and Aristotle proud. At Delicious, the relationships among the tags are much messier. For example, in a traditional tree, an object can be on only one branch. At Delicious, tagging a Web address with multiple tags in effect puts it on many branches. Yet despite the lack of a well-organized scheme of categories, Delicious can make a list of twenty thousand Web addresses thoroughly usable.

That was Schachter's first insight: Tags work as a way for individuals to remember and refind pages. His second was understanding the power of making people's lists public. At Delicious, you can not only find all the bookmarks you've tagged as "San Francisco," you can also find all the bookmarks anyone else at Delicious has tagged "San

Francisco"—or "San Francisco" and "restaurant," or you can add "vegetarian," "Chinese," and "cheap" to focus your search even more narrowly. Every time you check back at Delicious and click on a tag, you'll see the latest pages to which other people have applied the tag. It's like having a world of people with similar interests out scouring the Web for pages that you'll find interesting, relevant to your work, or simply delightful.

These *tag streams*—digitally assembled lists of pages that share a tag applied by people who may not even know one another—can be immensely useful if you need to follow the latest ideas and developments on a particular topic. If you're an industrial chemist, the tag stream of pages people have tagged "polymer" is likely to turn up information you would have otherwise missed. You could even subscribe to a *tag feed,* so that a daily list of new pages tagged "polymer" is automatically sent to your email in-box or to software—called an "aggregator"—designed to handle feeds. As you get into the habit, you may find yourself thinking that tagging a particular page "polystyrene" as well as "polymer" might help other chemists find and benefit from the page. Indeed, it's becoming common at technical conferences for the organizers to recommend that attendees tag their conference-related blog posts, photos, and online articles with a tag specific to that conference—"etech2006" or "poptech07"—so they can all be easily found by those using tag search sites such as Technorati .com. Because tagging is such an easy way to share knowledge, some companies, including IBM, are setting up their own internal Delicious sites so that research is shared within the company borders.

If you could take a top-down look at the tags at Delicious, you wouldn't see a tree. In fact, it would far more like the floor of a forest in autumn. There are millions of tagged bookmarks at Delicious, each with an average of two tags, and over half a million different tags. Printed out, those tags would be as orderly as confetti. But in the third order of order, the messiness of miscellaneous information doesn't reduce its utility. For example, users have uploaded over 225 million photos onto Flickr, the photo-sharing site—and are currently adding about 900,000 per day—and have applied 5.7 million different tags a total of 540 million times. Yet if you search at Flickr for photos tagged "Capri," it neatly divides them into photos of the island of Capri and of the Ford Capri by analyzing the tags people have applied. (Apparently, not enough people are photographing their pants for a pants cluster to emerge.) The clusters are surprisingly accurate given that they're based on nothing but the photos' tags. It turns out that the bigger the mess—more tags, and more tags per photo—the more accurate is Flickr's analysis. Other techniques are being developed for sorting the leaves into useful piles, including better photo recognition, bottom-up taxonomies, and even games—Google lets two strangers tag a photo simultaneously until they come up with the same word.

But we'll never be done making sense of these piles of information. Because tags are created by ordinary people using words that are meaningful to them, there will always be ambiguity. Is "SF" San Francisco, San Fernando, or Sally Field? That ambiguity can be a problem if you have to find absolutely every resource available. But if you're at Flickr to browse photos of San Francisco because you're planning to go as a tourist, it won't really matter if some of the more than 680,000 pictures tagged "San Francisco" are actually pictures of the San Francisco in Guatemala or if you miss a few thousand photos of the Golden Gate because they were tagged "SF." The ambiguity may even introduce us to other San Franciscos we want to visit.

Tagging is one way the miscellaneous is coming into its own, but it's not the only way. Objects that used to be organized by individuals or institutions are rapidly becoming available to us free of their old structures. Online music sites aggregate the world's music and let us access it in any order we want, as if it all resides on an unthinkably large jukebox. Wikipedia, the grassroots encyclopedia, does the same for encyclopedia entries. New online services let biologists refer to species without having to locate them in an often-contentious tree of life. While eBay turns the world of garage sales into a miscellaneous pile, Amazon does it for books, as does the University of Pennsylvania's PennTags project. The IBM consultant database does it for potential team members. What may be considered the twenty-first century's largest media company, Google, does it for Web pages. Dabble.com does it for videos. We are rapidly miscellanizing our world, breaking things out of their old organizational structures, and enabling individuals to sort and order them on the fly.

This goes far beyond simply organizing your information so you can find it again. It can change how a business works.

The British Broadcasting Corporation, known for years as "Auntie" because of its prim image, is tearing itself apart so that it can better accomplish its mission of bringing news and entertainment to British subjects. Since it began, in 1922, the BBC's content has been organized into programs, schedules, and channels. Today, the channels, like U.S. stations, are justifiably possessive of the shows over which they've labored. But as the millennium turned, the BBC realized that the ability to deliver radio (and eventually television) programs over the Internet meant the audience would no longer behave according to the BBC's schedule or way of organizing itself.

The system the BBC had to wrangle was a classic second-order monolith. Sarah Hayes, head of media asset management, and her crew work in a light-filled, airy space in the BBC's busy headquarters, managing access to goods in an industrial warehouse kept five miles away. File a request and just as soon as someone can schlep out to the right spot on the fifty-five miles of shelves, it'll be put on a shuttle van and delivered to you—the very definition of what "instantaneous" isn't. So in 1999 the BBC started spending approximately $100 million a year to preserve and to convert its archived material to modern, digital formats.

To take advantage of third-order means of finding information, the BBC began a long and complex process of turning its layout of stations and schedules into a miscellaneous pile of programs. This breakdown of the traditional ways of organizing content affects every aspect of the BBC's business, from how it compensates its channels to their licensing agreements with producers and artists who thought they were going to control when and how often the programs were going to "air." But the BBC realized that changing the rules of broadcasting enables their viewers to get more value from the content the BBC produces. People want to be able to listen to or watch programs whenever they want. When listeners are trying to find, say, a jazz performance, they don't care if the program originated on BBC Radio 1 or Radio 4. In the digital world, channels make more sense to the creators of the information than to the users of it. The audience can also find programs long buried in the BBC archives and watch them when they want to. And not only watch: The BBC has also been working on clearing the rights for programs so viewers can use portions to create new works of scholarship and creativity. It is slow and expensive work, and the BBC's progress has been uneven, in part due to changes in management But this radical "mixing it up" of programs—both by untethering them from their broadcast schedules and by making them available for reuse—sharply increases the BBC's value to its customers, which is precisely its mission, and a goal for every business.

MISCELLANEOUS FROM A TO Z

On paper, it sounds like a terrible idea. Build an encyclopedia by letting anyone create or edit an article, even anonymously. Yet four years after its launch at the beginning of 2001, Wikipedia had more people reading its pages than the *New York Times'* Web site did. By the middle of 2006, Wikipedia boasted over a million articles in its English edition, with more than a hundred editions in other languages. The traditional sources of authoritative knowledge have begun to pay close attention to the new kid on the block, and not only to its content. Traditionally, the articles in a work that size would be carefully arranged. But Wikipedia's organization is as bottom-up as its content.

The *Encyclopaedia Britannica* does not have the luxury of being as thoroughly miscellaneous as Wikipedia. If we're looking for the *Britannica*'s article on elephants, we count on being able to open the volume with the *E* stamped on its spine and page through alphabetized entries until we get to the one we want. If we're feeling adventurous, we can check out the carefully planned cross references at the end of the article. Or we can go to Mortimer Adler's *Propaedia* to find a family of Adler-approved concepts related to elephants. Either way, we are able to find information in the *Britannica* precisely because it isn't miscellaneous.

At Wikipedia, there are no volumes—not even digital representations of volumes—to thumb through. There is an alphabetical listing of the topics, but it's poorly done—Mortimer Jerome Adler is listed under the *M*'s—probably because the listing is rarely used. There are tens of thousands of entries for each letter, on average. That's a lot of riffling, whereas with eight keystrokes and a press of the Enter button, you could have searched for *elephant* and found the article about pachyderms instantly. At the top of the elephant article, there's a link to a page that lists all the other articles in Wikipedia you might have meant to find when you typed *elephant* into the search box: a film by Gus Van Sant, an album by the White Stripes, a World War II German antitank vehicle, a brand of beer, or the 105th chapter of the Koran. Wikipedia reminds us that even a word as simple as *elephant* has a touch of the miscellaneous about it.

Even if you use Wikipedia's alphabetical index, the pages are not really in alphabetical order. In fact, a Wikipedia article isn't a single object. Although an article's Web page looks unified to the reader, as with many pages on the Web, its text, graphics, and formatting rules are each stored separately and are pulled together only when a user requests a page by clicking on a link. If you search for *elephant* at the Wikipedia site, it's probably the computer named Vincent (after Vincent of Beauvais, a Dominican priest who compiled an encyclopedia with 3,718 chapters in the thirteenth century) that comes up with the list of articles that use the word. If you click on the link to the main article, this sends a request to another computer, which checks to see if that article was recently requested by someone else; if so, a copy of that page is kept ready to go and a third computer—perhaps the one named Will Durant, after the historian of philosophy—simply sends the page you're looking for. If not, Wikipedia sets about constructing the page for you. It randomly looks at one of the half dozen computers (including one named after Mortimer Adler) that store the complete text of the current articles in Wikipedia. Wikipedia then looks on Bacon (named after the philosopher Sir Francis Bacon) or one of the other computers that store the graphics, and passes both the text and the graphics to one of the dozens of computers that do nothing but assemble contents into Web pages based on templates. The finished page is then passed to your computer, where you see a text-and-graphics page about elephants.

Another level down, Wikipedia, like all computer applications, is even more miscellaneous. The computer may decide to store any single element of an article—say, the text or a photo of an elephant—in discontinuous sectors of a hard drive in order to fit the most data onto the drive and to optimize the time it takes to retrieve all those bits. That's why when I asked Brion Vibber, the chief technical officer of the Wikipedia organization, where the text information for the elephant article is actually stored, he replied, in the chat room we were in:

<brion> god only knows.
<brion> On the disk somewheres

A shame-faced admission of an appalling ignorance? Not at all. The gap between how we access information and how the computer accesses it is at the heart of the revolution in knowledge. Because computers store information in ways that have nothing to do with how we want it presented it to us, we are freed from having to organize the original information the way we eventually want to get at it. The bits and pieces of Wikipedia are, in effect, an enormous reserve of miscellaneous information that can be assembled in precisely the ways we need at precisely the moment we need it. That's true all the way through Wikipedia, from the microscopic bits stored on the hard drives to the finished articles we read.

At the top level of this hodgepodge of bits, images, text, articles, and ideas, something remarkable happens. The million articles in English are not arranged alphabetically. They are not put into a

Dewey-like categorization scheme. There is no controlled vocabulary. There is no usable overview. Yet this enormous miscellany gets organized richly and in tremendous detail. How it happens would have driven Mortimer Adler over the brink: Wikipedia articles are packed with hyperlinks created by anyone who takes the time to add one. No qualifications are required, and no expertise is needed beyond knowing that to link the word *elephant* in an article to its entry in Wikipedia, you type "[[elephant]]". In some entries, almost every second word is linked to another article. Together these links constitute a web of knowledge, communally constructed, ever shifting, and frequently extraordinarily useful.

Wikipedia's hyperlinked web, like the Web itself, does not look like a tree. It is a far, far more complex structure. But its shape, freed from the two dimensions of paper, better represents the wild diversity of human interests and insight.

NEW PROPERTIES, NEW STRATEGIES, NEW KNOWLEDGE

College students' silverware drawers, Delicious, Flickr, the BBC, and Wikipedia are miscellaneous in different ways, except for one thing: How their content is actually arranged does not determine how that content can and will be arranged by their users. In some cases—Wikipedia, for example—no one even knows exactly where the raw contents are. These examples are miscellaneous *because* users don't need to know the inner organization, *because* that inner order doesn't result in a preferred order of use, and *because* users have wide flexibility to order the pieces as they want, even and especially in unanticipated ways. This means that the miscellaneous enables *all* of the information contained in the set to be discovered over time.

But this also means the miscellaneous doesn't much resemble our traditional view of knowledge. Knowledge, we've thought, has four characteristics, two of them modeled on properties of reality and two on properties of political regimes.

As we've seen, the first characteristic of traditional knowledge is that just as there is one reality, there is one knowledge, the same for all. If two people have contradictory ideas about something factual, we think they can't both be right. This is because we've assumed knowledge is an accurate representation of reality, and the real world cannot be self-contradictory. We treat ideas that dispute this view of knowledge with disdain. We label them "relativism" and imagine them to be the devil's work, we sneer at them as "postmodern" and assume that it's just a bunch of French pseudointellectual gibberish, or we say "what*ever*" as a license to stop thinking.

Second, we've assumed that just as reality is not ambiguous, neither is knowledge. If something isn't clear to us, then we haven't understood it. We may not be 100 percent certain about whether the Nile or the Amazon is the longest river, but we're confident one is. Conversely, if there's no possibility of certainty—"Which tastes better, beets or radishes?"—we say it isn't a matter of knowledge at all.

Third, because knowledge is as big as reality, no one person can comprehend it. So we need people who will act as filters, using their education, experience, and clear thinking. We call them experts and we give them clipboards. They keep bad information away from us and provide us with the very best information.

Fourth, experts achieve their position by working their way up through social institutions. The people in these institutions are doing their best to be honest and helpful, but until humans achieve divinity, our organizations will inevitably be subject to corrupting influences. Which groups get funded can determine what a society believes, and funding is often granted by people who know less than the experts: The fate of a DNA research center may rest with congresspeople who can't tell a ribosome from a trombone.

The way we've organized knowledge has been largely determined by these four properties of knowledge. We've tried to settle on a single, comprehensive framework for knowledge, with categories so clear and comprehensive that experts can put each thing in its proper place. Institutions grew to maintain the knowledge framework. Their ability to certify experts and to vouch for knowledge made them powerful and, sometimes, rich. So when the miscellaneous shakes our certainty in the nature of knowledge, more than the future of the card catalog is at stake. Because a third-order miscellany is digital, not physical, we no longer have to agree on a single framework. Things have their *places,* not a single place. We get to create our own categories, ones that suit our way of thinking. Experts can be helpful, but in the age of the miscellaneous they and their institutions are no longer in charge of our ideas.

These are big changes, but perhaps the most urgent one is this: Over the course of the millennia, we've developed sophisticated methods and processes for developing, communicating, and preserving knowledge. We have major institutions—serious contributors to our culture and our economy—devoted to those tasks. We're good at it. Now we have to invent new ways appropriate to the new shape of knowledge. We are doing so at a pace unparalleled in our history.

Four new strategic principles are emerging, severing the ties between the way we organize physical objects and ideas.

Filter on the way out, not on the way in. A friend of mine who worked at the *Harvard Business Review* tells amusing stories about the "slush pile," the unsolicited manuscripts that arrive every day. The *Harvard Business Review* is a sober journal of research and ideas, yet people submit poetry, short stories, and arty photographs. My friend's job was to go through the slush pile to see what, if anything, was worth passing along for serious consideration. She was a gatekeeper, a filterer, doing a job that makes sense when the economics and physics of paper force us to make decisions about what knowledge we will publish and thus preserve. We rely on experts such as my friend to spare us from having to wade through the slush pile on our own.

But when anyone can publish at the press of a button, the social role of gatekeepers changes. For example, from the outside the "blogosphere" looks like a self-indulgent pool of slush that wouldn't get past the usual publishing filters. While the economics of publishing ensures that most blogs indeed wouldn't be let through the gates, the aggregate value of all the blogs in the "long tail" (to use the term Chris Anderson made popular in his book of that name)—each perhaps of interest to only a few people—is incalculable. This is an inversion of the old model. In a world of parsimonious access to paper, filters increase the value of what's available by excluding the slush. But in the third order, where there's an abundance of access to an abundance of resources, filtering on the way in *decreases* the value of that abundance by ruling out items that might be of great value to a few people. Filtering on the way out, on the other hand, increases the value of the abundance by locating what's of value to a particular person at a particular moment. For example, a physics professor at McGill University, Bob Rutledge, started an electronic bulletin board that posts new findings for *any* astronomy research as soon as it can be summarized. Rutledge doesn't apply criteria to decide for the reader whether the research is important enough to be included (though only active, professional astronomers can register to post to the site). It's up to each reader to be the filterer. Similarly, the Public Library of Science's biology journal, a peer-reviewed but free online resource, started PLoS One in November 2006. "The idea is to take the editorializing out of the peer review process," says Hemai Parthasarathy, the managing editor. So long as a paper is "sound," it will be published. If it's good science, *someone* may find it useful. So long as the user has good tools for finding what she needs—and this is a task many are working on—filtering on the way out vastly increases our shared potential for knowledge.

Put each leaf on as many branches as possible. In the real world, a leaf can hang from only one branch. In the first order of organization, there's no way around that limitation. In the second order, most cataloging systems have provisions for listing books under more than one heading, but the physicality of the second order still usually demands that one branch be picked as the primary one, and there is a limit on the number of secondary listings.

In the third order, however, it's to our advantage to hang information from as many branches as possible. If you get a new Casio digital camera to sell in your online store, you'll want to list it under as many categories as you can think of, including cameras, travel gear, Casio products, graduation gifts, new items, sale items, and perhaps even sports equipment. Hanging a leaf on multiple branches makes it more findable by customers. Unlike in the second order, this doesn't make your e-store disorganized or messy. It makes it more usable . . . and more profitable.

Everything is metadata and everything can be a label. In a store, it's easy to tell the labels from the goods they label, and in a library the books and their metadata are kept in separate rooms. But it's not so clear online. If you can't remember the name of one of Shakespeare's plays, go to the search box at Google Book, type "Shakespeare tragedy," and you'll see a list of all of them. Click on, say, *King Lear* and you can read the full text, including the famous line, "How sharper than a serpent's tooth it is to have a thankless child!" Now suppose you want to know where the quotation "How sharper than a serpent's tooth" comes from. Type the phrase into the search box and Google Book will list *King Lear.* Simple, but in the first case you used Shakespeare's name as metadata to find the contents of a book and in the second you used some of the contents of the book as metadata to find the author and title. In the miscellaneous order, the only distinction between metadata and data is that metadata is what you already know and data is what you're trying to find out.

In the first two orders of order, we've had to think carefully about which metadata we'll capture because the physical world limits the amount of metadata we can make available: A book's catalog card has to hold far less information than does the book itself. In the third order, not only can every word in a book count as metadata, so can any of the sources that link to the book. If we want to help our customers or users find information, we'll try to make as much of it usable as metadata as we can.

This not only makes sites easier to use, it vastly increases the leverage of knowledge. Think of what we can do with just the few words that fit on a second-order card or a label. Now that everything in the connected world can serve as metadata, knowledge is empowered beyond fathoming. We not only can find what we need based on whatever slight traces we have in our hand, we can see connections that would have escaped notice in the first two orders.

The power of the miscellaneous comes directly from the fact that in the third order, everything is connected and therefore everything is metadata.

Give up control. Build a tree and you surface information that might otherwise be hidden, just as Lamarck exposed information left hidden in Linnaeus's miscellaneous category of worms. But a big pile of miscellaneous information contains relationships beyond reckoning. No one person or group is going to be able to organize it in all the useful ways, hanging all the leaves on all the branches where they might be hung. For example, iTunes shows users a branch that pulls together albums by a particular artist, but the millions of playlists that users have made there find relationships that the organizers of iTunes could not possibly have foreseen, from techno versions of children's songs to tracks played at someone's third wedding. iTunes simply cannot predict what people are going to be interested in, what a song is going to mean to them, and what connections they're going to see. Some of the combinations will be of passing value to only one person, but other people may find their world changed by how a stranger has pulled together a set of songs to express a mood, an outlook, or an idea.

That's why it's so powerful to let users mix it up for themselves. Go into a real world clothing store and try pulling everything in your size off the racks and into a shopping cart so you can go through it in an orderly fashion. After all, that's the rational way to proceed. Everything that's not your size is just noise, a distraction. Yet, within ninety seconds you'll be thrown out of the store and firmly asked not to return. On line, on the other hand, we just naturally expect to organize information our way, through tags, bookmarks, playlists, and weblogs. And then we add to the information that a site provides us by disagreeing with it in our own reviews. Users are now in charge of the organization of the information they browse. Of course, the owners of that information may still want to offer a prebuilt categorization, but that is no longer the only—or best—one available. Put simply, the owners of information no longer own the organization of that information.

Control has already changed hands. The new rules of the information jungle are in effect, transforming the landscape in which we work, buy, learn, vote, and play.

6

SMART LEAVES

In 1948, two graduate students at the Drexel Institute of Technology in Philadelphia overheard the president of a local grocery chain asking a dean to sponsor research into how to read product information automatically. The students, Joseph Woodland and Bernard Silver, inspired by the dots and dashes of Morse code, came up with a set of straight lines much like the modern zebra-stripe bar codes, and in 1951 they unveiled a machine that could translate the bar codes back into numbers. It was the size of a desk, wrapped in black oilcloth, and used a 500-watt bulb as the light source. "It could cause eye damage," Woodland recalled.

In 1966—four years after Silver died, at the age of thirty-eight—the idea went commercial when the National Association of Food Chains put out a call for automatic checkout machines to speed up checkout lines. The first was an RCA system installed at a Kroger store in Cincinnati in 1972, but to get real efficiency, bar codes would have to be put on the packages by the manufacturers, not the clerks working in the local stores. So the association established the Uniform Grocery Product Code, the grandparent of the Universal Product Code (UPC) standard we use today. In 1974, at a Marsh Supermarket in Troy, Ohio, the first working system successfully identified a ten-pack of Wrigley's Juicy Fruit chewing gum that is now housed in the Smithsonian. In 1981, the U.S. Department of Defense required bar codes on all products it purchased and the UPC system went mainstream. Today there are about five billion items scanned every day, in more than 140 countries.

As we enter the third order of order, bar codes are providing a handy gateway between physical products and digital information about those products. At LibraryThing.com, where people share lists of books, if you upload a photo of a book's bar code, it will look up the information it needs to add an entry to your personal library. It will even automatically add an image of the front cover. PULP—Personal Ubiquitous Library Project—a felicitously named project sponsored in part by Microsoft, intends to let corporations build their own internal libraries in the same way. And once we have a unique identifier for a third-order object, it's poised to become a smart leaf, because now all the comments, metadata, and associations people make with that leaf become findable.

But there's a rub. It's clear to every clerk and customer that the bar code on the blister pack of six kitchen scrub pads applies to the blister pack overall; if the cashier scans it six times, the customer will rightfully complain that the cashier just doesn't get how bar codes work. In the second order, the manufacturer gets to declare what units it wants to track with a bar code; in the third, anyone with an Internet account can pull together ideas and information from anywhere she wants, extracting a single thought, word, or image out of the "blister pack" of an online resource. In the second order, the bar code gets stamped well after the manufacturer has decided what constitutes a product and how it'll be packaged. In the third order, stamping an ID on a leaf often is what turns it into a leaf in the first place.

Which means that before we can ask how we're going to connect the leaves to make them smart, we first have to figure out what the leaves are.

THE VALUE OF POINTING

"Fisticuffs almost broke out when latecomers could not gain entrance to an overflow session on UPC bar coding at the National Retail Merchants Association convention here Monday," reported *Women's Wear Daily.* It was 1987, the year UPCs surged to indisputable market acceptance. At the time of the conference there were already 25,000 manufacturers using the codes, but new industries, sensing a chance to vastly improve their efficiency, were getting ready to jump on board. The clothing industry had adopted UPCs as a voluntary standard in 1986. In 1989, the seafood industry would get the go-ahead to expand the number of UPC digits it used in order to accommodate seafood products sold by random weight, a move so successful that a few years later it expanded the number of digits again.

The UPC number not only lets checkout lines move faster, it makes the entire inventory-tracking process more efficient, and drives down a merchant's costs because it points to a bloom of information in the merchant's database. The merchant typically has to add only one more visible piece of data to prepare an item for sale: the price. In a 1986 study, 80 percent of businesses said that their UPC equipment paid for itself in less than two years, and 45 percent said the return on their investment took less than one year; a quarter found they were saving $100,000 a year (about $170,000 in current dollars) directly and the same amount indirectly. No wonder the UPC session at the National Retail Merchants Association convention was packed.

The digits of the Universal Product Code created a global system of information that has helped create a global system of commerce. But putting the "Universal" into the UPC requires some work. UPC codes consist of twelve numbers, although starting in 2005, retailers' scanners have had to accommodate the thirteen-digit European Article Number (EAN) as well. A UPC is divided into three parts: a manufacturer identifier, a product identifier, and a digit calculated from the other digits that serves as a check on the integrity of the number. The manufacturer number is assigned by the GS1, the group that owns the UPC system, but it's up to each manufacturer to come up with codes for their particular products. The question of what constitutes a product is settled by how manufacturers and merchants need to track items. A manufacturer of winter apparel will typically assign separate UPC numbers for the small, medium, and large versions of its snow hats, as well as separate numbers for each color they come in, even though in some sense they are the same product.

Manufacturers can assign any number to any item, without having to employ Deweyesque catalogers to figure out if they should classify snow hats under "winter equipment" or "outerwear." Any number will do. But that also means that within and across industries, the identifiers have no significance: The UPC number for a GE fluorescent bulb has no relationship to the UPC for a Sylvania fluorescent bulb, so inventory systems have no obvious way of seeing how many bulbs are sold overall.

The United Nations Standard Products and Services Code, developed in 1998, takes the opposite approach. Its code numbers are full of meaning, representing the branches on a gigantic five-level tree of stuff that classifies everything from cats ("Live Plant and Animal Material and Accessories and Supplies>Live animals>Livestock") to voting rights defense associations ("Organizations and Clubs>Civic organizations and associations and movements>Human rights advocacy and defense associations"). To reduce the friction of commerce, the two systems are being integrated: In 2003 the UNSPSC handed over management of its system to the group that manages UPCs. But the merger can't be complete because UPCs are used primarily for goods packaged for sale, while the UN numbers are also used for moving raw materials around the world. The world is so diverse in its things and how we view them that we're never going to agree entirely; even when the intentions are the best and the leadership is unified, there will also be a miscellaneous residue.

UPCs are a success story. But they are an old technology, created in the 1970s, when computers were the size of a Broadway star's changing room. Now tiny Radio Frequency Identification (RFID) tags broadcast information about the products they're attached to. Manufacturers are less happy about RFIDs than they were about bar codes, however. Bar codes can be printed on an item's box, while RFID tags need to be purchased, programmed, and affixed. But RFIDs can contain more information than UPCs and can be integrated directly with computing systems. RFIDs are already being used in the automatic lanes of tollbooths, to tag cows, to detect U.S. Energy Department prohibited materials, and to track all of the cargo and equipment used in the Iraq War. Kroger estimates that RFID tags attached to temperature sensors could cut spoilage in half, saving the chain hundreds of millions of dollars a year. A University of Arkansas study commissioned by Wal-Mart showed that out-of-stock items made it to the shelves three times faster if they had RFID tags. Three Virginia hospitals are using RFIDs to track ten thousand movable pieces of health-care equipment so that they can get more use out of less inventory. The hospitals expect the system to pay for itself in just a year. Those sorts of numbers—especially as the cost of the tags comes down—will drive businesses to tag their packages with RFIDs, literally embedding digital metadata into their physical systems. Physical objects may even become "spime"—science-fiction writer Bruce Sterling's name for a physical object whose location and status is constantly tracked and recorded—becoming smarter not only for the benefit of the user but also for the manufacturer, who gets a perfect record of what customers are doing with it.

The bar code goes on the cereal box and the RFID is clipped to the orange snow hat. Intellectual content—packages of ideas and information—is not as easy to pin down. The BBC is discovering the benefits of identifying fine-grained elements of its inventory.

In addition to the BBC's gargantuan library of television and radio programming of over one million hours, the BBC archives hold more than four million photos, 1.2 million CDs and vinyl albums, and four million pieces of sheet music for the BBC's five orchestras. If every one of these items is carefully labeled with its title, subject, and date of airing, it still will not help a producer find and reuse fifteen seconds of footage of a Beijing traffic jam that's part of a documentary on globalization; the producer is going to have to pay for a new shoot. Treating a program as a smart leaf doesn't automatically enable the *elements* of a program to be treated as smart leaves. The solution is to create even more metadata. One initiative, shepherded by

John Good and Carol Owens, identified a standard set of metadata to add to the BBC's materials. Pointing to the seven to eight hundred production units that contribute content to the BBC, Good notes the most obvious point: "All the systems need to know what 'title' means." Standard labels such as "title" ensure that catalogers don't frustrate searchers by using labels such as "name" or "program title." Once the labels are agreed on, the data has to be put into the fields in a standardized way, so that one episode's title isn't listed as "Monty Python: Episode 11" and another's as "Monty Python's Flying Circus #12." Otherwise, the computing system may not be able to identify all the episodes in the series. The BBC system standardizes over three hundred different attributes that may apply to recorded material, including subject, producers, language, length, type of media, even whether it has won any awards. Good says it is used to tag anything from an entire series to a single frame in a particular country's version of a particular program.

A second initiative, begun in 2003 under Tom Coates and Matt Webb, looked at "how people find programming, how they navigate around it, how they use it." Coates and Webb kept coming back to one basic problem: Even with the BBC's new standardized system, there wasn't agreement about what exactly the identifier should point at. A series? A season? A particular translation? A series of broadcasts, with separate numbers for rebroadcasts? Coates and Webb ultimately decided that the most useful object—the one that accords best with how the audience thinks about programming—was an *episode,* perhaps one particular hour of a TV series, or a certain broadcast of the eight-to-ten P.M. jazz show. Hence in addition to the tags, each episode is assigned a unique five-character identifier, as meaningless as a UPC, which is used as the end of the address of a Web site automatically created for it. But Coates and Webb understood that someone searching for Episode 11 of *Monty Python* might well want the rest of the shows, too. So their system automatically creates a Web address for the entire series, with the episodes as pages underneath them, much as a site might have addresses such as www.usa.com/florida/miami.html and www

.usa.com/florida/orlando.html. The treelike structure of the program sites captures important information about the relationships among the programs.

In deciding which elements will get an identifier, each of the two systems makes choices about what constitutes a useful, identifiable leaf. The identifiers provide a "hook" on which can be hung all the metadata required to let users filter the contents for themselves. With such a system in place, producers will be able to reuse segments as small as a frame and viewers will be able to find entire series. Both systems, together and apart, will bring the BBC's key assets new value just by letting them be pointed at smartly.

INCLUDE AND POSTPONE

When the seafood industry adopted standard UPCs, it also converted its terminology to match the U.S. Food and Drug Administration's names for over two hundred species of fish and shellfish, first published in 1988 as *The Fish List: FDA Guide to Acceptable Market Names for Food Fish Sold in Interstate Commerce.* (In 1993 the FDA ruined the near rhyme by renaming it *The Seafood List.*) Look up *bonito* in the online version and you'll find seventeen fish going by that name, from the amberjack (*Seriola dumerili*) to the northern bluefin tuna (*Thunnus tonggol*). At the top and bottom of the page, the Office of Seafood warns: "NOTE: Use of the vernacular name is not encouraged, and may cause the seafood to be misbranded."

Scientists, not just fishmongers, have problems branding fish. The fish known as "bonito" on the East Coast of the United States is classified as *Sarda sarda,* but the fish known as "bonito" on the West Coast is *Sarda chiliensis.* And if a West Coast scientist comes to the East Coast and catches a *Sarda sarda,* she'll exclaim, "Wow, look at the size of that skipjack!" In fact, there are almost a hundred different scientific and common names for bonitos, not just the seventeen itemized in *The Seafood List.*

David Remsen, the chief librarian at the Marine Biology Laboratory in Woods Hole, Massachusetts, recognized that the widespread disagreement about what marine organisms are called meant that scientists searching his library weren't finding all the information about any particular fish. Remsen couldn't wait until all questions about what to call species were resolved, especially since it seems they never will be. So he began the uBio project, or the Universal Biological Indexer and Organizer—an attempt to "take taxonomy out of the eighteenth century" by creating a comprehensive and collaborative catalog of the names of all living (and once-living) organisms.

It is an entirely third-order idea, and it is no small job. Into one pile goes every name the project can find for each of the 1.5 to 1.75 million species of plants and animals on the planet; Remsen estimates that between scientific and common names, there are 10 million overall. Instead of trying to limit the number of names an organism has, he welcomes any and all possibilities; no matter how regional or oddball, it's better to have it in the pile than out. Into a second pile goes every reasonable scientific taxonomy he can find. So far, he has identified sixty different taxonomies, and he hopes to at least double that in the next few years.

It's a perfect example of the wisdom of the basic two-pronged strategy for going miscellaneous: *Include and postpone.* On the one hand, uBio includes every name it can find, even the bonito's slang name in the Canary Islands, because someone sometime might find it useful to know what the name refers to. On the other hand, uBio also postpones when the classifying and organizing occurs because scientists don't agree. Rather than building the system around David Remsen's idea of what the right taxonomy is, uBio can accommodate many more scientists by letting them make their own decisions. Remson designed uBio not to deliver the single right answer but to provide the maximum *potential* knowledge.

Getting a handle on species is such a fundamental need that there are several efforts in addition to uBio's. The All Species Foundation has set itself the goal of recording every name of every species, and of describing them all, within twenty-five years. The International Commission on Zoological Nomenclature has created an online register, ZooBank, for species names. Unlike uBio, though, it wants to be the official gatekeeper, deciding which names are the acceptable ones. The Life Sciences Identifier (LSID) project—sponsored by IBM, Oracle, and Sun, among others—provides much more latitude, enabling organizations that already have life science databases, including medical and pharmaceutical research firms, to share information without having to change anything in their databases. An LSID consists of six parameters separated by colons—similar in form to Ranganathan's Colon Classification system—that uniquely identify the data so that researchers can refer to it unambiguously. The Tree of Life Web project, founded in 1993, is building its own collection, giving each species its own page and linking all of them into a single tree. More than 320 biologists have contributed to it.

The proliferation of solutions may appear to be making matters worse, but the old system, while it seemed straightforward, actually masked some serious issues, as you can see back at the Linnean Society headquarters in London. Inside, carefully preserved, are biological specimens hundreds of years old that look like little more than carefully shaped dust. The specimens are the ultimate *what* to which Linnean names point. Without the specimens, there would be no way to settle arguments about exactly which species a name points to. But you can't expect to settle all disputes about biological classification by slapping an ancient fish down on the table and saying, "Now *that's* a *Sarda sarda,* buddy!" Someone else can always slap a slightly different specimen down and claim that it's of the same species. Is it sufficiently different that it ought to be counted as a separate species? That's an argument biologists have been having for hundreds of years, but it's indicative of a far more widespread dispute that arises as soon as people start to identify things, especially when there are lots of things and lots of people. How are we ever going to coalesce far-flung knowledge about something—a species, a book, a TV episode, an orange snow hat—if we can't agree what that thing is?

THE ESSENCE OF THE MATTER

Birders in southwestern Africa are atwitter over all the new species emerging in field guides, including the Barlow's lark, the southern black korhaan, and the agulhas long-billed lark. These are not new discoveries, like the long-tailed pipit; they are new splits among birds that were formerly lumped together. At issue is not merely where to stick a particular South African lark with brown feathers and white-and-black trim. Rather, we are plunged directly into the question of what constitutes a species, a concept that seems to have gotten harder to pin down the more our culture has thought about it.

It was much more straightforward so long as we believed in *essentialism*—the idea that everything is defined by clear and knowable traits that make it into what it is. For Aristotle and the tradition he spawned, the essence of a species is the set of characteristics that uniquely define it as a category. Linnaeus began his career with the standard Christian belief (heavily influenced by Aristotle) that God populated Eden with the various species, each then exactly as it is now. But observing how interbreeding creates hybrids, Linnaeus came to believe that the rungs of the Great Chain of Being were filled in over time as animals bred themselves into diversity, exactly according to God's precise plan.

Modern biologists are more like "accidentalists" than essentialists. The course of evolution depends on which animals mount which others and what mutations occur that enhance their ability to produce robust offspring. Nor do biologists today think there's a system of perfectly distinct essential characteristics that define species perfectly. There are too many ways species can be considered similar or different. That's why your position in the controversy over the long-billed lark (*Certhilauda curvirostris*) depends on which traits—color of the plumage, shape of the bill, similarity of DNA, where they live, with what birds they breed, the songs they sing in flight—you count as determinative.

Since Darwin considered species to be the very things that evolution acts on, you would think he would have offered a clear definition of the term. Yet two years before *Origin of Species* was published, Darwin wrote:

> It is really laughable to see what different ideas are prominent in various naturalists' minds, when they speak of "species"; in some, resemblance is everything and descent of little weight—in some, resemblance seems to go for nothing, and Creation the reigning idea—in some, sterility an unfailing test, with others it is not worth a farthing. It all comes, I believe, from trying to define the indefinable.

And in *Origin of Species* he writes that "we shall have to treat species . . . as artificial combinations made for convenience" in order to be free from the "vain search for the undiscovered and undiscoverable essence of the term species." Given the title of his book, we can be confident that Darwin did not mean to say that species are merely fictitious or arbitrary ways of carving up the animal kingdom. Darwin was pointing to the difficulty of defining *perfectly*—note his swipe at essentialism in the phrase "essence of the term species"—what for him was a very real joint of nature. But biologists have argued ever since Darwin over where exactly to carve. One expert, Marc Ereshefsky, counts a dozen different concepts. Some scientists argue quite seriously that species denote nothing real, a position held by none less than Thomas Jefferson, who argued that "Classes, orders, genera, species, are not of her [Nature's] work. Her creation is of individuals."

So, arguments over the long-billed lark go on. Having lost essentialism, we don't have a replacement that does as good a job at divvying up the things of the world. We don't even have confidence that there *is* an inarguable way to divide the world into types of things. And that's a problem, because as the world becomes more miscellaneous, if we can't pin something down, we can't coalesce information around it.

Essentialism is still with us, even if it is no longer an option for evolutionary scientists. As the philosopher James Danaher points out, we think there is a disease called "cancer" with some essential traits.

The cure, we've assumed, will attack those essential traits. But cancer now seems to be a collection of hundreds of diseases. What we call breast cancer alone may include dozens of different diseases, each of which may have multiple causes. The same may be true of bipolar disorder, schizophrenia, the midlife crisis, and the common cold. The search for *a* cure for each of these is a result of our essentialism.

Businesses suffer from the effects of essentialism, as well, when they assume they know what their products are for—are you sure your company's "energy bar" is being eaten to help with a workout and not as a candy?—and when they stick too closely and too long to their mission statements. The very concept of a market smacks of essentialism's tendency to define matters too clearly. Marketers have acted as if their job is to come up with messages that will appeal to markets segmented by demographics. Particular demographic properties are selected because the marketers believe they define a group susceptible to the same message; hence 18-to-24-year-old suburban males get the "It'll make you tough" advertising while the females get the "Boys will like you" message. The market exists as something that will receive a message. Stop "messaging" it and the "market" of 18-to-24-year-old suburban males exists only as one of a nearly infinite number of ways of slicing and dicing a population. Thinking that 18-to-24-year-old suburban males exist as a market—as something more than a random slice—gets in the way of seeing the truly fascinating phenomenon: miscellaneous customers finding one another in the digital world and forming real social groups, not because they share essential demographic traits but because they're talking with one another. The markets that conversations make are real markets, not mere statistical clusterings.

Essentialism makes the world seem more manageable, but it can lead us to miss what's really going on.

WHAT IS A BOOK?

There's something comforting about the sight of cards spooning in a library card catalog. A world of ideas and knowledge, more than we could ever absorb, is waiting for us, carefully indexed in those neat rows of drawers. And yet the second order masks a complexity the third order confronts head-on: We don't really know what a book is.

We think we do because the second order of order reduces information so drastically. That's how it works: Card catalogs have value because of what they leave out. Melvil Dewey himself designed the current standard card in 1877, sizing it at 7.5 by 12.5 centimeters, roughly 3 by 5 inches, very close to the golden rectangle so prized by the ancient Greeks and Renaissance artists. Because it's not very large, catalogers have to make tough decisions about what information to include. Generally you'll find the book's call number, author, title, publisher, place of publication, date of publication, number of pages, size, International Standard Book Number (ISBN), subject heading, and whether it's illustrated. Generally you will *not* find how well the book sold, if it's been banned in any countries, a list of the books it cites, the college the author attended, what the reviewers said about it, the full index from the back of the book, or how many times it's been checked out of the library—although, as John Seely Brown and Paul Duguid point out in *The Social Life of Information,* you can sometimes tell if a card has been heavily consulted by how dog-eared it is.

But what are you going to do? A library card is a second-order object, so catalogers make the best decisions they can. Card catalogs enable us to navigate the library by giving us only a narrow slit to look through. The digital world, on the other hand, has never met a piece of information it didn't like—and couldn't put to work. It just needs a handhold—such as the International Standard Book Numbers publishers have used since the 1960s to identify each edition of every book they sell.

The ISBN of Herman Melville's *Moby-Dick* illustrated by Rockwell Kent is 0679600108. At the Library of Congress site, a search of that ISBN reveals that the book is a Modern Library edition, 822 pages long, 21 centimeters high, printed on recycled, acid-free paper. At Amazon.com, a search of the ISBN connects us to Amazon's analysis of the book's distinctive phrases ("pagan harpooners"), the fact that yesterday this edition was the 43,631th most bought book but today it's fallen to 49,581, that it contains 208,968 words, that its Fog index (a standard measure of readability) says it's of medium difficulty, that your purchase gets you 14,634 words per dollar, and that 286 people have written reviews—every one of which you can read—and have awarded it an average of four out of five stars. You can also go to ISBN.nu, set up in 1998 by the journalist Glenn Fleishman, to get information about where to buy the book online and a list of the various editions available under other ISBNs, including audio versions. At LibraryLookup.com—created by Jon Udell, another journalist—you can enter the ISBN to see if your local library has a copy of the book. The PULP project will pull together information about the book from multiple sites, including reviews and annotations. At Harvard's experimental H20 site, you can find all the registered courses that have *Moby-Dick* on their syllabi, including an MIT course called "Major Authors: Melville and [Toni] Morrison," suggesting a connection most of us would not have made.

Smart leaves are not like catalog cards with more room and an extra forty IQ points. Rather than having a dollop of information contained in a small rectangle, an endless web of information sprawls across the indefinite space of the Web. An identifier such as an ISBN that enables distributed information to come together when needed turns a C-student leaf into a genius.

But ISBNs, like bar codes, are fundamentally second-order constructs, specifying precise editions of books. Unless you happen to be a publisher or the owner of a bookstore, you're interested in arguing about *The Da Vinci Code*, not the paperback edition or the large print edition of it. We don't have a system of identifiers for books regardless of editions because the world of intellectual content is way more complex than it usually seems.

What is *Hamlet*? The correct *Jeopardy!* question is, of course, "What is a play by Shakespeare, Alex?" But if your professor tells you to read *Hamlet* for tomorrow's quiz, you can't run out and buy the play, a phrase that taken literally doesn't make sense. You'll instead look for a *copy* of the book. However, it's a copy with no original.

comparison of the texts isn't as easy as it seems either, since not only may books have different prefatory materials, editions may have extensive footnotes or facing-page annotations. Humans can mentally bracket those additions, but computers aren't as nimble as we are. A standard text comparison would also report that *Hamlet* with its original spellings and *Hamlet* in a typical modern version are definitely two different books; *Hamlet* in French, Spanish, Swahili, and Turkish would count separately, too. The OCLC database of books, like a card catalog, doesn't have the full text of books, in any case. Hickey's project, xISBN, augments its automated comparisons with human insight, relying on librarians to hand-cluster books. Since the ISBNs say nothing about how any particular book is related to any other, xISBN has to rely on clever programming, human intervention, and guesswork. As a result, if you key in an ISBN for *Hamlet,* "you'll get a long list of *Hamlet* editions," says Hickey. Will you get all of them? No. And you may even get some results that aren't *Hamlet.* But gathering leaves imperfectly is better than leaving them scattered.

Librarians understand that between the Platonic idea of *Hamlet* and the tattered paperback copy carried around by a high school student, books exist on several planes. So the International Federation of Libraries Association created the Functional Requirements for Bibliographic Records (FRBR) standard. The most abstract concept it describes is a *work,* such as *Hamlet* in all the different ways it is performed and published. Next it defines an *expression,* such as the First Folio version or the Folger edition of the work. Then there is a *manifestation,* which puts the expression onto paper (or CD, or the Internet), such as the Folger hardcover, paperback, and large-print editions. Finally there is the *item*—actual copies of the book. All this sounds quite neat, but it gets messy quickly. Is the version of *Hamlet* rewritten for children with a happy ending still *Hamlet*? How about works inspired by *Hamlet,* such as Tom Stoppard's *Rosencrantz and Guildenstern Are Dead* and Lisa Fiedler's *Dating Hamlet: Ophelia's Story*? The FRBR says that when the modification of a work "involves a significant degree of independent intellectual or artistic effort," it becomes a new work. That seems reasonable (if we can agree on what constitutes a "significant degree"), but we still want to know that there's a relationship between Folger's complete *Hamlet* and Sesame Street's *Bert and Ernie Meet Hamlet.* This is one of the virtues of a tree: If all these editions and versions were carefully placed on branches, we could walk up the tree to discover that they're all varieties of *Hamlet.*

But *Hamlet* is too complex for trees. We have a remarkable vocabulary for talking about bookish things. There are abridged versions, translations, annotated versions, modern renderings, books based on others, side-by-side translations, plot summaries, graphic novels, poetry collections inspired by, reinterpretations, study guides, audiobooks, modern-language versions, and parodies. Various sets of these relationships hold for various expressions and manifestations of the work. No tree can express these without becoming more exception than rule.

There's complexity in another dimension as well. *Hamlet* consists of acts and scenes, at least since the First Folio. The scenes consist of numbered lines, but also of speeches divided up by characters. In some editions, those lines and speeches have those pesky footnotes, another level of granularity. *Hamlet* itself is a part of a larger collection known as Shakespeare's tragedies, and the tragedies are part of Shakespeare's plays, part of his works, part of Elizabethan literature. What constitutes a leaf in this new vision of a miscellanized world? The play? A scene? A famous quotation from the play? A line? All of a character's lines? Elizabethan plays? The deeper we look, the less the leaf metaphor holds up. Leaves imply entities that are well defined and knowable, that have edges and a persistence sufficient that we can count them or at least point at them. Instead, we're staring at a pile of leaves that all seem to have *something* to do with *Hamlet,* even though we can't find an easy way to identify them as such and can't even find a way to identify the thing they're all related to.

It's enough to make you long for essentialism. But the third order doesn't let us become strict constructionists who recognize only a narrow range of essential *Hamlet* works and variants. Leaves are smart because of the unpredictable, open-ended ways we are able to relate them. We can't come up with one thing we can definitively call *Hamlet* because books and the works they record are so rich and complex. The range of relationships is as broad as the human imagination. As we mix computers and human intelligence to rake together all sorts of leaves about books and their relationships, those leaves will be associated in more and more ways, perpetually building our miscellaneous pile of leaves in fits and starts. It will be imperfect because there is no one ideal *Hamlet* we can blister-pack and bar-code once and for all, but that imperfection is also a source of richness.

For example, at some site, someone will see the phrase "More honored in the breach than the observance" and will tag it "*Hamlet*" and perhaps even "act 1, scene 4, line 16." And then the quotation on that site will have become a smart leaf. Its tags will link it with an online facsimile of the First Folio of the play. That facsimile is a leaf, and the association is a leaf as well. A journal article about act IV, scene 3, becomes a leaf that through a tag or analysis of the text can be associated with *Hamlet*. So does an editorial cartoon in a high school newspaper showing the school principal dressed like Hamlet. The connections between these pieces are potential, waiting to be found and used. Every tag, every link, every computer sweep through the online world enriches our potential for seeing connections and understanding things in contexts we had never considered.

This web of information, knowledge, insight, and opinion is possible only because none of these leaves has a place defined as clearly and sharply as an Aristotelian species. Its value comes from the jumble, from the fact that its ordering is postponed. And of course this holds not just for books and classic plays but for the entire world of third-order information, with consequences beyond how we organize ideas. In health care, for example, your physician traditionally was not only the expert, she was the only expert you were allowed to see, even though sometimes she'd give you special permission for a "consultation" with another, approved, medical expert. It took a few years of widespread Internet access for the medical profession to get used to the idea that while we would continue to grant our doctor sole authority over our treatment, we were no longer willing to let her be the sole source of information. The medical leaves are already too connected for that. If you are diagnosed with diabetes, you will very likely find yourself browsing the Web, trying to understand how the disease will affect your life. Or maybe you just want to see if grapefruit jolts blood sugar levels. The page you find will have links. The links will lead you into official sites from accredited experts, such as Diabetes.org, the home of the American Diabetes Association, but the links will also take you on an unpredictable tour of blogs by diabetics, scientists, and crackpots, and discussion boards where diabetics talk about the daily particularities of life with the disease: Why does my blood sugar increase after exercising? Has anyone else been craving black coffee? Where can I get sugar-free cheesecake? That information doesn't fit in the *Physicians' Desk Reference*. It doesn't fit in the head of even the most expert of doctors. It exists only because it was created, one leaf at a time, by a world of people with their own interests. It comes together only when someone needs it to. It's there as a potential only because diabetes as a topic is as loose, multifaceted, and broadly entangled as *Hamlet*.

INTERTWINGULARITY

"People keep pretending they can make things deeply hierarchical, categorizable, and sequential when they can't. Everything is deeply intertwingled." So said Ted Nelson, the eccentric visionary who coined the term *hypertext* in the mid-1960s. In the third order of order, information not only becomes intertwingled, intertwingularity enables knowledge. And unique identifiers enable intertwingularity—although there can be so many unique identifiers for the same thing and at various levels of abstraction that the identifiers are all a-twingle also.

Microsoft Research—the research-and-development group that Microsoft sponsors—has been working for years on a project it calls "AURA," Advanced User Resource Annotation. If you use your cell phone, or other device, to take a picture of a bar code on a product, AURA connects to the Internet to find out everything that's known about that product. Marc Smith, the head of the research group, describes a trip to the grocery store. When he scanned his favorite breakfast food, AURA found a headline stating that the FDA had recalled the cereal because its ingredients list was inaccurate. "Annotation" is part of the AURA acronym—"Annotate the planet!" is its slogan—because it allows users not just to find information about a product but also to rate and comment on it, associating yet more information with it for the world at large.

In the intertwingled world of the near future, let's say Smith runs his AURA-enabled phone over the bar code of a can of tuna. The program will serve up mercury warnings and recipes for *salade niçoise* from the Internet. If he's unfamiliar with *salade niçoise,* a click could take him to Wikipedia or to a history of food from the Riviera. He might also find the four entries for tuna at the U.S. FDA Regulatory Fish Encyclopedia: albacore, kawakawa, skipjack, and yellowfin. If Smith clicks on the skipjack entry, he'll find himself in the Scombridae family of seven mackerels and tunas. Number three is *Sarda chiliensis,* the Pacific bonito, where he can review photographs, geographic information, its distinctive pattern of proteins, and—eventually—its DNA sequence. (The DNA bar-coding project, which uses snippets of DNA as unique identifiers for species, will enable lots of leaves to be pulled together.) Smith could then use the official scientific name *(Sarda chiliensis)* or its ordinary name (Pacific bonito) to look it up at uBio. There he could learn the fish's other common names, which would let him Google his way through scientific papers, cookbooks, and folktales. He could get linked to Amazon's collection of books about bonitos, including *Scombrids of the World: An Annotated and Illustrated Catalogue of Tunas, Mackerels, Bonitos, and Related Species Known to Date,* ISBN 925101381-0. That ISBN could lead him Lord knows where. In this web of intertwingled resources, ground-breaking research on medical applications of the bonito family are one link away from someone's vacation story about fishing in the Pacific and just a few links away from the greatest whaling story ever told, perhaps in the edition illustrated by Rockwell Kent.

In the next aisle, Smith could focus his cell phone's camera on the bar code on a bottle of dish soap and discover that it irritates some people with particular allergies and that it's handy for drowning ticks. The bar code on a baby's squeeze toy could be two clicks from advice on how to travel on a plane with an infant or an argument over whether Bach or the Beatles is better music for walking a baby back to sleep.

But unique identifiers don't just provide a way to pull information together. They also allow information to be dispersed. That's why Ulla-Maaria Mutanen, a Finn who writes the HobbyPrincess blog, created Thinglinks.org, a Web service that assigns unique IDs to the items craftspeople create. Unlike an ISBN, a Thinglink denotes an individual item because each handmade object is unique. For example, in April 2006, the University of Art and Design in Helsinki issued a Thinglink ID for each item in its exhibit of work by graduating master's students. That way people can write about the works on their blogs, on review sites, at online craft markets, or anywhere they want. Because each ID is distinctive—THING:378RGD, for example—search engines are able to locate every Web page that mentions it, whereas a blog post that refers only to "a lovely brown wool sweater I saw the other day" will get lost in the nearly infinite shuffle of leaves. The distinctiveness of the ID makes it possible to decentralize the discussion of the sweater with the confidence that it will all remain intertwingled.

The identifiers in this stew are themselves mixed. Some are as carefully assigned as Thinglinks, bar codes, ISBNs, and uBio identifiers. Others are as loose as the vernacular names for a fish. When it's possible to identify leaves clearly and cleanly, the unique identifiers can enable an extraordinary distributed development of related ideas, making the individual leaves smarter and smarter. In such cases, meaningless IDs do a better job of postponing the ordering of the miscellaneous. And every piece, component, and particle should be ID'd because someday someone will want to refer to just that one bit. But where IDs don't make sense, we'll still connect every idea we can.

It may be harder for our computers to assemble all the leaves that talk about something as loosely defined as *Hamlet* or diabetes, but we're only going to get better at this. We have to. It's how we're going to make sense of the miscellany of ideas and information we're creating for ourselves.

SOCIAL KNOWING

Every day, an iconic scene plays out in newsrooms around the country. Middle-aged white men sit around a table in a room with windows opening onto a vast, fluorescent-lit work space filled with desks and busy, busy people. It is the daily editorial meeting made famous in *All the President's Men* and dozens of other films and television shows. In this instance Hollywood gets it right. Editorial meetings are a pinnacle of power at newspapers. If you work hard as a journalist for many years, you just may be invited into the club.

The editors exert their power by deciding what to build on one of the most valuable pieces of real estate in the world: the two square feet or so that are the front page of the newspaper. Through their choices, the editors tell us what they think were the most significant events of the previous day. They rank each story through a code readers implicitly understand: Where on the page is it? Is it above the fold? How big is the headline? Did it merit a byline? Does it have a cute subhead to draw the reader in? Editors count on our being able to read the page's body language.

Digg.com, which describes itself as a "user driven social content website," also has a front page. It's not particularly pretty, featuring a playlist of headlines with two lines of summary. Next to each headline is a number representing the number of "diggs"—readers' thumbs-ups—each article has received. Any reader can suggest a story, and if enough people then vote the story out of the "Digg area queue," it gets its time on the front page. The discussion about why a story is important takes place not around a table in an interior room but in public pages accessed through links on the front page that lead to comments left by dozens and sometimes hundreds of readers who talk with one another about the story's accuracy, importance, and meaning.

Digg is hardly unique. A similar site, Reddit.com, was created by a couple of college students over their summer vacation. Common Times is a Digg for left-of-center politicos. Similar sites are springing up every couple of weeks because they regularly turn up articles worth reading. In fact, the idea of using readers as an editorial board is already expanding in two useful directions: Sites that have nothing to do with news are using it, and sites are arising that determine the rankings based on social groups within the general readership. For example, if you tell Rojo.com who your friends are among other Rojo users, it will list stories that are popular among them. You can mark particular stories as of likely interest to your friends so that when they next visit Rojo, it will show them a list of stories you've recommended, including your comments. Another site, TailRank, lets you "narrow down your results to just news from your feeds, your tags, and your buddies," says Kevin Burton, its president. Reddit is adding a similar capability. Rollyo.com searches many different types of sources—not just news—working off lists supplied by friends and celebrities, so you can see, for instance, what's for sale in the set of online shopping sites compiled by your friends plus celebrities such as Debra Messing and Diane von Furstenberg.

Not all of these sites will survive. Indeed, some are likely to have vanished in the time it takes to bring this book to print. But some will survive and others will arise, because enabling groups of readers to influence one another's front pages not only brings us more relevant information, it also binds groups socially.

This binding is certainly different from the way broadcast media have formed one nation, under Walter Cronkite. With everyone seeing the same national news and reading the same handful of local newspapers, there was a shared experience that we could count on. Now, as our social networks create third-order front pages unique to our group's interests, we at least get past the oft-heard objection that what Nicholas Negroponte called *The Daily Me* fragments our culture into isolated individuals. In fact, we are more likely to be reading *The Daily Me, My Friends, and Some Folks I Respect.* We're not being atomized. We're molecularizing, forming groups that create a local culture. What's happening falls between the expertise of the men in the editorial boardroom and the "wisdom of crowds." It is the wisdom of groups, employing *social expertise,* by which the connections among people help guide what the group learns and knows.

The *New York Times* was founded in 1851 and the Associated Press in 1848. Such organizations have a resilience that should not be underestimated. But they will need it if they are to survive the ecological change that is occurring. We simply don't know what will emerge to challenge newspapers, any more than Melvil Dewey could have predicted Google or the *Britannica* could have predicted Wikipedia.

Dollars to donuts, though, the change will be toward the miscellaneous, and it will draw on social expertise rather than rely on men in a well-lit room.

THE CONUNDRUM OF CONTROL

In February 2005, Michael Gorman, the president of the American Library Association, lambasted weblogs in the association's flagship magazine, *Library Journal:*

> A blog is a species of interactive electronic diary by means of which the unpublishable, untrammeled by editors or the rules of grammar, can communicate their thoughts via the web. . . .
>
> Given the quality of the writing in the blogs I have seen, I doubt that many of the Blog People are in the habit of sustained reading of complex texts. It is entirely possible that their intellectual needs are met by an accumulation of random facts and paragraphs.

Some librarians—especially those who were also Blog People—were outraged. "An example of irresponsible leadership at its worst," wrote Sarah Houghton on her blog, Librarian in Black. "Excoriating ad hominem attacks wrapped in academic overspeak," blogged Free Range Librarian Karen Schneider, adding, "No citations, of course." The best title of a blog post had to be "Turkey ALA King" by Michael D. Bates at BatesLine.

Gorman brushed off his critics, citing his "old fashioned belief that, if one wishes to air one's views and be taken seriously, one should go through the publishing/editing process." How fortunate for Gorman that he heads an organization with its own journal.

But then Gorman is hardly alone in his skepticism about online sources. In October of the same year, Philip Bradley, a librarian and Internet consultant, said in the *Guardian* that Wikipedia is theoretically "a lovely idea," but "I wouldn't use it, and I'm not aware of a single librarian who would."

Robert McHenry, a former editor in chief of the *Encyclopaedia Britannica,* summed up his analysis of Wikipedia:

> The user who visits Wikipedia to learn about some subject, to confirm some matter of fact, is rather in the position of a visitor to a public restroom. It may be obviously dirty, so that he knows to exercise great care, or it may seem fairly clean, so that he may be lulled into a false sense of security. What he certainly does not know is who has used the facilities before him.

If these experts of the second order sound a bit hysterical, it is understandable. The change they're facing from the miscellaneous is deep and real. Authorities have long filtered and organized information for us, protecting us from what isn't worth our time and helping us find what we need to give our beliefs a sturdy foundation. But with the miscellaneous, it's all available to us, unfiltered.

More is at stake than how we're going to organize our libraries. Businesses have traditionally owned not only their information assets but also the organization of that information. For some, their business *is* the organization of information. The Online Computer Library Center bought the Dewey Decimal Classification system in

1988 as part of its acquisition of Forest Press. To protect its trademark, in 2003 the OCLC sued a New York City hotel with a library theme for denoting its rooms with Dewey numbers. Westlaw makes a good profit providing the standard numbering of court cases, applying proprietary metadata to material in the public domain. But just about every industry that creates or distributes content—ideas, information, or creativity in any form—exerts control over how that content is organized. The front page of the newspaper, the selection of movies playing at your local theater, the order of publicly available facts in an almanac, the layout of a music store, and the order of marching bands in the Macy's Thanksgiving Day Parade all bring significant value to the companies that control them.

This creates a conundrum for businesses as they enter the digital order. If they don't allow their users to structure information for themselves, they'll lose their patrons. If they do allow patrons to structure information for themselves, the organizations will lose much of their authority, power, and control.

The paradox is already resolving itself. Customers, patrons, users, and citizens are not waiting for permission to take control of finding and organizing information. And we're doing it not just as individuals. Knowledge—its content and its organization—is becoming a social act.

ANONYMOUS AUTHORS

The real estate industry maintains its grip on its market through the National Association of Realtors' control of the nation's 880 local multiple listing services (MLS). NAR is North America's largest trade association, the third-largest lobby, and was the third-largest donor in the 2004 presidential election. It has almost 1.3 million members, which means that one out of every 230 Americans belongs to it. NAR protects its members' interests by locking low-cost brokers out of local listings, defending the standard 5 to 6 percent fee. So when real estate sites like PropSmart.com and Zillow.com came along, NAR wasn't happy. PropSmart automatically scours the Web, populating

Google maps with every real estate listing it can find. If a user finds a listing she's interested in at PropSmart, the site puts her in touch with the local real estate agent offering it. Although this would seem to be nothing but a benefit for the local agents, Ron Hornbaker, the founder of PropSmart, regularly receives angry letters from MLS lawyers because, with fees for residential property reaching $61 billion in 2004, NAR is desperate to keep the listing centralized and under its control. As PropSmart and Zillow add features that allow users to sort through listings by distance from schools, environmental quality, and crime safety statistics—pulling together leaves from multiple sources—NAR is right to feel that its business model is being threatened. The threat comes not from particular sites such as PropSmart but from the difficulty of keeping information from becoming miscellaneous.

The miscellanizing of information endangers some of our most well-established institutions, especially those that get their authority directly from their grip on knowledge. The *Encyclopaedia Britannica* is up-front about where its authority comes from, writing that its editorial board of advisors includes "Nobel laureates and Pulitzer Prize winners, the leading scholars, writers, artists, public servants, and activists who are at the top of their fields." The *Britannica* trumpets past contributors such as Albert Einstein, Sigmund Freud, and Marie Curie. The credibility of its authors and editors is the bedrock of the *Britannica*'s authority.

No wonder Wikipedia took the *Britannica* by surprise. Wikipedia has no official editors, no well-regulated editorial process, no controls on when an article is judged to be ready for publication. Its authors need not have any credentials at all. In fact, the authors don't even have to have a name. Wikipedia's embrace of miscellaneous, anonymous authorship engenders resistance so strong that it sometimes gets in the way of understanding.

How else to explain the harsh reaction to the now famous "Seigenthaler Affair"? For four months in the spring of 2005, Wikipedia readers could find an article that matter-of-factly claimed that the respected print journalist and editor John Seigenthaler was implicated in the assassinations of both John and Robert Kennedy. It was a particularly vicious lie given that Seigenthaler had worked for Robert Kennedy and was a pallbearer at his funeral.

As soon as Seigenthaler told a friend about it, the friend corrected the article. But Seigenthaler was shocked that for those four months, anyone who looked him up would have read the calumny. "At age 78, I thought I was beyond surprise or hurt at anything negative said about me. I was wrong," he wrote in an op-ed for *USA Today*. "Naturally, I want to unmask my 'biographer.' And, I am interested in letting many people know that Wikipedia is a flawed and irresponsible research tool."

As Jimmy Wales, the founder of Wikipedia, later said in response to the media hubbub: "Wikipedia contained an error. How shocking!" Wales was mocking the media, not downplaying Seigenthaler's distress. Indeed, Wales responded quickly with a change in Wikipedia's ground rules. No longer would anonymous users be allowed to initiate new articles, although they could continue to edit existing ones. The media headlines crowed that Wikipedia had finally admitted that real knowledge comes from credentialed experts who take responsibility for what they say. Wikipedia was growing up, the media implied.

Unfortunately, in their eagerness to chide Wikipedia, the media got the story backward and possibly inside out. In fact, Wales's change *increased* the anonymity of Wikipedia. Registering at Wikipedia requires making up a nickname—a pseudonym—and a password. That's all Wikipedia knows about its registered users, and it has no way to identify them further. If you don't register, however, Wikipedia notes your Internet protocol address, a unique identifier assigned by your Internet service provider. Requiring people to register before creating new articles actually makes contributors more anonymous, not less. Explains Wales, "We care about pseudoidentity, not identity. The fact that a certain user has a persistent pseudoidentity over time allows us to gauge the quality of that user without having any idea of who it really is." At Wikipedia, credibility isn't about an author's credentials; it's about an author's contributions.

The *Encyclopaedia Britannica* has Nobelists and scholars, but Wikipedia has Zocky. Zocky has contributed mightily in hundreds of articles. If Wales and the Wikipedia community see that an edit was made by Zocky, they know it has value because Zocky's many contributions have shaped a reputation. But no one—including Wales—knows anything else about her or him. Zocky could be a seventy-year-old Oxford don or a Dumpster-diving crack addict. The personal peccadilloes of the greatest contributors to Wikipedia might top those of the greatest citizen contributor to the *Oxford English Dictionary,* the famously criminally insane murderer who did his research from his prison hospital. Would it matter?

To succeed as a Wikipedia author—for your contribution to persist and for it to burnish your reputation—it's not enough to know your stuff. You also have to know how to play well with others. If you walk off in a huff the first time someone edits your prose, you won't have any more effect on the article. You need to be able to stick around, argue for your position, and negotiate the wording. The aim is not to come up with an article that is as bland as the minutes of a meeting; the article about Robert Kennedy, for instance, is rather touching in its straightforward account of the reaction to his assassination. Wikipedia insists that authors talk and negotiate because it's deadly serious about achieving a neutral point of view.

Neutrality is a tough term. The existentialism of the 1950s, the New Journalism of the 1960s, and contemporary postmodernism have all told us not only that humans can't ever be neutral but also that the claim of neutrality is frequently a weapon institutions use to maintain their position of power and privilege. The first time I talked with Jimmy Wales, I started to ask him about the impossibility of neutrality. Wales politely and quickly cut me off. "I'm not all that interested in French philosophy," he said. "An article is neutral when people have stopped changing it."

This is a brilliant operational definition of neutrality, one that makes it a function of social interaction, not a quality of writing to be judged from on high. And Wikipedia's approach usually works quite well. Take the entry on the Swift Boat Veterans for Truth, the group that during the 2004 presidential campaign attacked John Kerry's Vietnam war record. Let's say you're incensed that the Wikipedia article states that Kerry earned three combat medals. From your point of view, he may have earned one, but the other two were awarded inappropriately and, in any case, had nothing to do with combat. So you edit the reference to call them "service" medals. When you later return to the article, you notice that someone has reinstated the word "combat." You could reinstate your change, but it's likely that it will be changed back to "combat." You could go to the "discussion" page attached to each Wikipedia article and explain why you think calling them "combat medals" wrongly tilts the page in favor of Kerry. Or you could come up with alternative language for the article that you think would satisfy those who disagree with you; for example, you could add details describing the controversy over those medals in a manner all disputants would accept. Either way, Wikipedia edges closer to the neutral point of view so valued by Wikipedians that they've turned it into the acronym NPOV.

This very controversy arose at the Swift Boat article. On the article's "discussion" page you can read the back-and-forth, one contributor declaring that he must "absolutely oppose" the use of the word "combat," others arguing that "combat" is the right word. The conversation turns angry. One contributor throws up her or his virtual hands, addressing another's "poisoned" behavior. At other times the participants work together to come up with wording that will meet everyone's idea of what is fair and accurate. Trying out shades of meaning in a process that can look like hairsplitting—did Kerry toss "decoration items" or "decoration paraphernalia" over the White House fence?—the discussion is actually a negotiation zeroing in on neutrality.

Of course, not everyone who reads the Swift Boat article is going to agree that NPOV has been achieved. But if you question the neutrality of the Swift Boat article, you'd be well-advised to read the discussion page before making an edit. In a high-visibility article such as this one, you are likely to find that the point that bothers you has been well discussed, evidence has been adduced, tempers have cooled, and language has been carefully worked out. On rare occasions when agreement can't be reached and the page is being edited back and forth at head-whipping velocity, Wikipedia temporarily locks the page—but only temporarily.

It may take years for a discussion to settle down. Samuel Klein, who describes himself as a "free knowledge activist" and is the director of content for the One Laptop Per Child project in Cambridge, Massachusetts, is a respected contributor to Wikipedia under the pseudonym SJ. He describes an argument that raged for three years about whether articles that mention the sixth-largest city in Poland should refer to it as Gdańsk, as it's called in Polish, or Danzig, as it's known in German. The "edit war" was so ferocious that it was finally put to a vote, which determined that when referring to the city between 1308 and 1945, articles should use its German name, but the Polish name for any other period. The vote also decided that at the first use of the name, the other name should be placed in parentheses. If you want to read the arguments and follow the evidence, it's all there in the discussion pages of Wikipedia, open to anyone. (Imagine if we could read the discussion pages about a 1950s Wikipedia entry on segregation.)

Wikipedia works as well as it does—the journal *Nature*'s discovery that science articles in Wikipedia and *Britannica* are roughly equivalent in their accuracy has been a Rorschach test of the project—because Wikipedia is to a large degree the product of a community, not just of disconnected individuals. Despite the mainstream media's insistence, it is not a purely bottom-up encyclopedia and was never intended to be. Wales is a pragmatic Libertarian. A former options trader, in 2000 he cofounded an online encyclopedia called Nupedia that relied on experts and peer review; it was funded mainly by the money Wales made as a founder of Bomis, a "guy-oriented" search engine that knew where to find soft-core porn. Over the course of three years, twenty-four articles had completed the review process at Nupedia. Expertise was slowing the project down. He and a colleague founded Wikipedia in 2001, and Wales left Nupedia in 2002.

Wales, who is both fierce about his beliefs and disarmingly non-defensive about them, emphasizes that from the beginning his aim has been to create a world-class encyclopedia, not to conduct an experiment in social equality. When the quality of the encyclopedia requires sacrificing the purity of its bottom-upness, Wales chooses quality. Yet the media is continually shocked to discover that Wikipedia is not purely egalitarian, trumpeting this as if it were Wikipedia's dirty secret. Wales is proud of the fact that a community of about eight hundred people has emerged to curate and administer Wikipedia as needed. These administrators are granted special privileges: undoing a vandal's work by reverting pages to previous versions, freezing pages that are rapidly flipping back and forth in an edit war, even banning a contributor because he repeatedly restored contested edits without explaining why. This type of hierarchy may be anathema to bottom-up purists, but without it, Wikipedia would not work. Indeed, in the "stump speech" he gives frequently, Wales cites research that shows that half of the edits are done by just less than 1 percent of all users (about six hundred people) and the most active 2 percent of users (about fifteen hundred people) have done nearly three-quarters of all the edits. Far from hiding this hierarchy, Wales is possibly overstating it. Aaron Swartz, a Wikipedia administrator, analyzed how many letters were typed by each person making edits and concluded that the bulk of *substantial* content is indeed created by occasional unregistered contributors, while the 2 percent generally tweak the format of entries and word usage. In either case, Wikipedia is not as purely bottom-up as the media keeps insisting it's supposed to be. It's a pragmatic utopian community that begins with a minimum of structure, out of which emerge social structures as needed. By watching it, we can see which of the accoutrements of traditional knowledge are mere trappings and which inhere in knowledge's nature.

And what is the most important lesson Wikipedia teaches us? That Wikipedia is possible. A miscellaneous collection of anonymous and pseudonymous authors can precipitate knowledge.

AUTHORITY AND TRUTH

It wasn't enough for the Wizard of Oz to tell the truth. He had to tell the truth in an amplified voice that emerged from an amplified image of his visage, in a chamber grand enough to intimidate even the bravest of lions. Given a choice between truth and authority, the Wizard would probably have chosen the latter.

Social knowledge takes a different tack. When its social processes don't result in a neutral article, Wikipedia resorts to a notice at the top of the page:

> The neutrality of this article is disputed.
> Please see discussion on the talk page.

Wikipedia has an arsenal of such notices, including:

- The neutrality and factual accuracy of this article are disputed.
- The truthfulness of this article has been questioned. It is believed that some or all of its contents *might* constitute a hoax.
- An editor has expressed a concern that the topic of this article may be unencyclopedic.
- Some of the information in this article or section has not been verified and might not be reliable. It should be checked for inaccuracies and modified as needed, citing sources.
- The current version of this article or section reads like an advertisement.
- The current version of this article or section reads like a sermon.
- The neutrality of this article or section may be compromised by "weasel words."
- This article is a frequent source of heated debate. Please try to keep a cool head when responding to comments on this talk page.

These labels, oddly enough, add to Wikipedia's credibility. We can see for ourselves that Wikipedia isn't so interested in pretending it's perfect that it will cover up its weaknesses.

Why, then, is it so hard to imagine seeing the equivalent disclaimers in traditional newspapers or encyclopedias? Newspaper stories do sometimes qualify the reliability of their sources—"according to a source close to the official but who was not actually at the meeting and whose story is disputed by other unnamed sources who were present"—but the stories themselves are presented as nothing less than rock solid. And, of course, there are the small boxes on inner pages correcting errors on the front pages, ombudsmen who nip at the hands that feed them, and letters to the editor carefully selected by the editors. Yet the impression remains that the traditional sources are embarrassed by corrections. Wikipedia, on the other hand, only progresses by being up-front about errors and omissions. It Socratically revels in being corrected.

By announcing weaknesses without hesitation, Wikipedia simultaneously gives up on being an Oz-like authority and helps us better decide what to believe. A similar delaminating of authority and knowledge would have serious consequences for traditional sources of information because their economic value rests on us believing them. The more authoritative they are, the greater their perceived value. Besides, fixing an error in second-order publications is a much bigger deal because it requires starting up an editorial process, printing presses, and delivery vans. At Wikipedia, a libel in an article about a respected journalist can be corrected within seconds of someone noticing—and because so many leaves are connected, it can literally take seconds for someone to notice. When *Nature* magazine released its comparison of the error rates in Wikipedia and the *Encyclopaedia Britannica,* Wikipedians took pride in making most of the corrections to their entries within days and all of them within thirty-five days, though some of the changes, such as whether Mendeleev was the thirteenth or fourteenth child of seventeen, required extensive research. Wikipedia even has a page listing errors in the *Britannica* that have been corrected in the equivalent Wikipedia articles. There is no explicit indication of gloating.

Anonymous authors. No editors. No special privileges for experts. Signs plastering articles detailing the ways they fall short. All the disagreements about each article posted in public. Easy access to all the previous drafts—including highlighting of the specific changes. No one who can certify that an article is done and ready. It would seem that Wikipedia does everything in its power to avoid being an authority, yet that seems only to increase its authority—a paradox that indicates an important change in the nature of authority itself.

Wikipedia and *Britannica* derive their authority from different sources. The mere fact that an article is in the *Britannica* means we should probably believe it because we know it's gone through extensive editorial review. But that an article appears in Wikipedia does not mean it's credible. After all, you might happen to hit the article right after some anonymous wacko wrote that John Seigenthaler was implicated in the assassination of Robert Kennedy. And yet we do—reasonably—rely on articles in Wikipedia. There are other indications available to us: Is it so minor an article that few have worked it over? Are there obvious signs of a lack of NPOV? Is it badly written and organized? Are there any notes on the discussion page? Does it cohere with what we know of the world? These marks aren't that far removed from the ones that lead us to trust another person in conversation: What's her tone of voice? Does it sound like her views have been tempered by conversation, or is she dogmatically shouting her unwavering opinions at us? We rely on this type of contextual metadata in conversation, and it only occasionally steers us wrong. An article in Wikipedia is more likely to be right than wrong, just as a sentence said to you by another person is more likely to be the truth than a lie.

The trust we place in the *Britannica* enables us to be passive knowers: You merely have to look a topic up to find out about it. But Wikipedia provides the metadata surrounding an article—edits, discussions, warnings, links to other edits by the contributors—because it expects the reader to be *actively* involved, alert to the signs. This burden comes straight from the nature of the miscellaneous itself.

Give us a *Britannica* article, written by experts who filter and weigh the evidence for us, and we can absorb it passively. But set us loose in a pile of leaves so large that we can't see its boundaries and we'll need more and more metadata to play in to find our way. Deciding what to believe is now our burden. It always was, but in the paper-order world where publishing was so expensive that we needed people to be filterers, it was easier to think our passivity was an inevitable part of learning; we thought knowledge just worked that way.

Increasingly we're rejecting the traditional assumption of passivity. For ten years now, customers have been demanding that sites get past the controlled presentation of "brochureware." They want to get the complete specifications, read unfiltered customer reviews, and write their own reviews—good or bad. The Web site for the movie *The Da Vinci Code* made a point of inviting *anyone* to discuss the religious controversy of the film; by doing so, the studio reaped media attention, market buzz, and audience engagement. Citizens are starting not to excuse political candidates who have Web sites that do nothing but throw virtual confetti. They want to be able to explore politicians' platforms, and they reward candidates with unbounded enthusiasm when the candidates trust their supporters to talk openly about them on their sites.

In a miscellaneous world, an Oz-like authority that speaks in a single voice with unshakable confidence is a blowhard. Authority now comes from enabling us inescapably fallible creatures to explore the differences among us, together.

SOCIAL KNOWERS

Imagine two people editing and reediting a Wikipedia article, articulating their differences on the article's discussion page. They edge toward an article acceptable to both of them through a public negotiation of knowledge and come to a resolution. Yet the page they've negotiated may not represent either person's point of view precisely. The knowing happened not in either one's brain but in their conversation. The knowledge exists between the contributors. It is knowledge that has no knower. Social knowing changes *who* does the knowing and *how*, more than it changes the *what* of knowledge.

Now poke your head into a classroom toward the end of the school year. In Massachusetts, where I live, you're statistically likely to see students with their heads bowed, using No. 2 pencils to fill in examinations mandated by the Massachusetts Comprehensive Assessment System. Fulfilling the mandate of the federal No Child Left Behind Act, the MCAS measures how well schools are teaching the standardized curricula the state has formulated and whether students are qualified for high school degrees. Starting with the third grade, students' education is now geared toward those moments every year when the law requires that they sit by themselves and answer questions on a piece of paper. The implicit lesson is unmistakable: Knowing is something done by individuals. It is something that happens inside your brain. The mark of knowing is being able to fill in a paper with the right answers. Knowledge could not get any less social. In fact, in those circumstances when knowledge is social we call it cheating.

Nor could the disconnect get much wider between the official state view of education and how our children are learning. In most American households, the computer on which students do their homework is likely to be connected to the Net. Even if their teachers let them use only approved sources on the Web, chances are good that any particular student, including your son or daughter, has four or five instant-messaging sessions open as he or she does homework. They have their friends with them as they learn. In between chitchat about the latest alliances and factions among their social set, they are comparing answers, asking for help on tough questions, and complaining. Our children are doing their homework socially, even though they're being graded and tested as if they're doing their work in isolation booths. But in the digital order, their approach is appropriate: Memorizing facts is often now a skill more relevant to quiz shows than to life.

One thing is for sure: When our kids become teachers, they're not going to be administering tests to students sitting in a neat grid of separated desks with the shades drawn.

Businesses have long suffered from a similar disconnect. Businesses want their employees to be as smart and well informed as possible, but most are structured to reward individuals for being smarter and better informed than others. For example, at the Central Intelligence Agency, intelligence analysts are promoted based on the reports they write about their area of expertise. While there is some informal collaboration, the report comes out under the name of a single expert. There is no record of the conversations that shaped it. Not only does this diminish the incentive for collaboration, it misses the opportunity to provide the expanded context that Wikipedia's discussion pages make available. The CIA is hardly unusual in its approach. It's the natural process if the output consists of printed reports. Printing requires documents to be declared to be finished at some point, which tends to squeeze the ambiguity out of them. And, of course, printed documents can't be easily linked, so they have to stand alone, stripped of the full breadth and depth of their sources. But some in the CIA have become aware that these limitations can now be overcome: Blogs are providing useful places for floating ideas before they're ready to be committed to paper, and Intellipedia, an internal site using the same software as Wikipedia, has five thousand articles of interest to the intelligence community.

One of the lessons of Wikipedia is that conversation improves expertise by exposing weaknesses, introducing new viewpoints, and pushing ideas into accessible form. These advantages are driving the increasing use of wikis—online pages anyone can edit—within business. The CIO of the investment bank Dresdner Kleinwort Wasserstein, J. P. Rangaswami, found that wikis reduced emails about projects by 75 percent and halved meeting times. Suzanne Stein of Nokia's Insight & Foresight says "group knowledge evolves" on wikis. Michelin China began using a wiki in 2001 to share project information within the team and among other employees. Within three years, the wiki had four hundred registered users and had grown to sixteen hundred pages. Disney, SAP, and some major pharmaceutical companies are all using wikis.

If wikis get people on the same page, weblogs distribute conversation—and knowledge—across space and time. The mainstream media at first mistook blogs for self-published op-eds. If you looked at blogs individually, it's a fair comparison. With over 50 million known blogs (with 2.3 billion links), and the number increasing every minute, blogs represent the miscellaneousness of ideas and opinions in full flower. But the blogosphere taken as a whole has a different shape. Not only will you find every shading on just about any topic you can imagine, but blogs are in conversation with one another. So if you were interested in, say, exploring the topic of immigration, you could look it up in the *Britannica* or Wikipedia. Or you could go to a blog search engine such as Technorati, where you would find 623,933 blog posts that use that word and 38,075 that have tagged themselves with it. The links from each blog, and the commenters who respond to each blog, capture a global dialogue of people with different backgrounds and assumptions but a shared interest.

What you learn isn't prefiltered and approved, sitting on a shelf, waiting to be consumed. Some of the information is astonishingly wrong, sometimes maliciously. Some contains truths expressed so clumsily that they can be missed if your morning coffee is wearing off. The knowledge exists in the connections and in the gaps; it requires active engagement. Each person arrives through a stream of clicks that cannot be anticipated. As people communicate online, that conversation becomes part of a lively, significant, public digital knowledge—rather than chatting for one moment with a small group of friends and colleagues, every person potentially has access to a global audience. Taken together, that conversation also creates a mode of knowing we've never had before. Like subjectivity, it is rooted in individual standpoints and passions, which endows the bits with authenticity. But at the same time, these diverse viewpoints help us get past the biases of individuals, just as Wikipedia's negotiations move articles toward NPOV. There has always been a plenitude of personal points of view in our world. Now, though, those POVs are talking with one another, and we can not only listen, we can participate.

For 2,500 years, we've been told that knowing is our species' destiny and its calling. Now we can see for ourselves that knowledge isn't in our heads: It is between us. It emerges from public and social thought and it stays there, because social knowing, like the global conversations that give rise to it, is never finished.

8

WHAT NOTHING SAYS

Some labels are so dumb they're famous:

> On a Sears hair dryer: "Do not use while sleeping."
>
> On the packaging for a Rowenta iron: "Do not iron clothes on body."
>
> On a Nytol sleep aid: "Warning: May cause drowsiness."
>
> On Sainsbury's peanuts: "Warning: Contains nuts."
>
> On a child's cape costume: "Wearing of this garment does not enable you to fly."

Then there's our son's favorite: One day when he was eleven, I bought one of those self-igniting logs intended for suburbanites building fires for mood rather than warmth. To light it, you take a match to the paper wrapper itself, making a flame hot enough to ignite the chemical-soaked pressed-fiber "log" within. Proving the dominance of lawyers, the wrapper on the self-igniting log warns users that the product carries a risk of fire.

And my own favorite: In the mid-1970s, to illustrate the give-and-take in the latest round of nuclear arms negotiations, CBS News aired an animated graphic of a chessboard with missiles for pieces. The off-screen reporter explained that the Soviets had agreed to remove this many pieces, and the United States had agreed to remove that many. At the bottom of the screen, CBS News prominently displayed the label "Simulation" for those viewers who might otherwise have thought that the Americans and Soviets kept their missiles on a giant chessboard.

So imagine you were the proverbial Martian visiting the earth. You saw not only labels so dumb that we make fun of them on the Internet, but labels on fruit, labels on escalators to tell us where they end, escalators that *read* those labels out loud, and labels required by law to inform us that some other label was required by law. What conclusion could you reach except that you had stumbled upon the stupidest sentient species in the universe? "They have a relatively advanced consciousness," your report will say, "but they apparently have no understanding of the items they have made for themselves." Then you will arch one eyebrow and mutter, "Curious."

Usually, we're good at metadata. If you're one of the 82 million people who read *Parade* on Sundays, a headline in 2005 blaring, "Do You Have Diabetes?" might have caught your eye. After reading a few sentences, you probably checked the top of the page because something didn't feel right about the article. Sure enough, in small print you found the label "ADVERTISEMENT." You probably felt a little cheated because the layout, the typeface, and the tone of the headline all implicitly said the ad was an article. You were fooled by metadata.

But you were fooled not because we're bad at recognizing metadata. On the contrary, the *Parade* ad used our fluency with metadata against us. We are surprisingly subtle readers of metadata. No one taught us how to read a magazine cover, but we parse the front page of *Parade* with great delicacy. The size and position of the word "Parade" tells us that it's the name of the magazine, not the title of the lead article. We immediately understand that the name of our local paper printed just above "Parade" doesn't tell us anything about the content of the magazine, although the words "The World's Worst Dictators" do. We grasp all of this without any of it being explicitly labeled, because it's obvious from the implicit cues. We can read metadata before we learn to read.

We are so good at metadata that even its absence can be metadata. Consider what happens when we remove the spaces between words, as we must when creating Web addresses. The canonical example is

LumberjacksExchange.com, which in the usual all-lowercase format is indistinguishable from LumberjackSexchange.com. That site is no longer up. Nor is PenisLand.com (PenIsland). But TheRapistFinder .com (TherapistFinder) is. Using spaces to separate words now seems like such an obvious idea that it's hard to imagine why it took until 700 C.E. to come up with it. In part it's because writing was thought of as transcribing speech and only the robots among us pause after each word when speaking. Now we depend on the absence of letters, just as we rely on the absence of a smile to tell us that a speaker is serious about what she said.

THE IMPLICIT ECOSYSTEM

We marked the Earth first by wearing paths into it. One person found a way to get from A to B, and others followed it. As they walked, they wore away the vegetation. What started as a faint path became clearer as more people walked it, and as it became clearer, more people walked it. You didn't need a sign because the path was itself visible, a line worn into the earth. Paths are the original bottom-up, emergent phenomena.

One day, there were so many paths that we needed a marker to tell us that this one goes to the sandy cliffs and that one goes to the field of gray rocks. Before you know it, it's 1914, cars are running on roads that used to be paths, and the Automobile Club of Southern California decides it needs to mark the way from Kansas City to Los Angeles with four thousand signs. The implicit is made very explicit.

The path from implicit to explicit is not a one-way street. You can see the back-and-forth in the process by which new highway signs are developed and deployed. In the United States, for a sign to be accepted, the Federal Highway Administration has to agree to put it into the official Manual on Uniform Traffic Control Devices, a nine-hundred-page volume that lists each sign and every regulation about their use. There you'll learn that the BRIDGE ICES BEFORE ROAD sign may be removed or covered during seasons of the year when its message is not relevant. You can page wistfully through the illustrations of the familiar yellow diamond signs that depict bicycles, people, deer, cows, and other members of the category of Things That Cross Roads. You can feed your misery by thumbing through Discouraging Signs: NO OUTLET, DEAD END, and PAVEMENT ENDS. You'll even find the metasign that warns NO TRAFFIC SIGNS.

For the FHA to admit a new sign into the manual, the sign's symbols have to be clear. Not all make it. The FHA decided that people just wouldn't be able to make sense of a proposed sign depicting the little circle-headed man familiar from WALK signs fleeing saw-toothed waves of increasing size—a tsunami warning.

Although a sign's picture has to convey the message, to educate the public a sign will often initially have text as well. As the meaning of the graphic becomes second nature to the public, the text is removed. There used to be signs that read SIGNAL AHEAD. Today we just have the yellow diamond sign with a vertical rectangle containing red, yellow, and green circles.

The text is important information. Why remove it? In order to help the sign's meaning become implicit. Reading a sign takes longer than "getting" the symbol. When a symbol has become a part of our vocabulary, we don't stumble and fumble as we try to understand it. It's simply a part of the meaning of the world. It enters our context, our background. The irony is that traffic signs are there to make explicit what we otherwise might not notice: A signal is ahead so we should begin braking, or a tsunami may hit us so we should . . . well, it's not clear what that sign would expect us to do other than run like hell. A well-designed traffic sign makes us aware enough that we act appropriately, but not so aware that we have to think about it. We even know without thinking that an arrow pointing up means we should go straight forward, but one pointing down does not mean we should back up. We're that good at understanding the implicit.

Usually.

In 2006, someone posted an observation to Digg.com:

> WalMart online shoppers who buy the Planet of the Apes DVD set are shown similar items available for purchase. Among these are

documentaries on the life of Martin Luther King, Jr., boxer Jack Johnson, and Tina Turner. I am not sure how these items are similar, but it sure is offensive.

Someone using the pseudonym Nork wrote, "They've clearly classified this DVD set as black or african american themed. Way to go Wal Mart." "Possibly a lone website worker playing a sick joke?" said viperdaimao (0). "Submitter is a moron," chimed in odweaver. "They choose that based on what other people have went to from that." Fname had a slightly more detailed explanation:

> This is not a Wal*Mart problem, it's a problem in general of recommendation engines. Some white supremacists or racists probably went about viewing African-American themed movies, then browsed ape-themed movies, in order to get this result. Amazon has been bothered by this type of activity in the past.

In fact, a few months later, Amazon got a black eye because its automated system was telling people who were searching for books about "abortion" that they might actually have meant to search for "adoption." Amazon said that its software made the suggestion because the two words have similar spellings and because people who searched for one frequently searched for the other. But it was someone using the name "avantretard" who came up with the most likely solution to the Wal-Mart conundrum:

> Racism is a THEME explored in all those works. This combined with a subpar database system=NON STORY.

Until avantretard discovered the probable explanation, Wal-Mart's explicit recommendations lacked the implicitness required to understand them. By making the context explicit, the story became just an example of how wrong we can go when we attempt to let the explicit stand free of the implicit. That's one reason Amazon now provides a link to inform readers of what led it to make its recommendations.

It's even converted the link into a shopping feature, "The Page You Made."

Implicit context is fragile. It is easily lost as culture moves on. For example, no less than John Dryden and Alexander Pope thought a speech in the second act of *Hamlet*—it begins "The rugged Pyrrhus, he whose sable arms, / Black as his purpose, did the night resemble"—was so bad that they wondered if Shakespeare really was the author. Only in the late eighteenth century did someone propose what seems to be the correct interpretation: Shakespeare was trying to sound old-fashioned. To playgoers at the time, the speech would have stood out from the implicit context of everyday speech. A hundred years later, the meaning was lost because the implicit context was lost.

At a gas station in Allston, a community on the fringes of Boston, a do-it-yourself pump was giving customers trouble. They'd swipe their credit cards, choose an octane grade, and then stand there, stumped. The start "button" was a yellow rectangle with the word START written on it, not a raised button, and customers thought it was a label. Not knowing what to press to get the pump started, they'd get angry at being made to feel stupid. So the station posted a homemade sign pointing to the start button.

PRESS
START
BUTTON
< < < < <

Then people got more confused. They pressed the word "START" on the paper label so many times that they wore a thumb-sized hole straight through it. The start button had the metadata typical of a label, so they didn't press it. Since they couldn't find anything that had the metadata of a button, they pressed the paper START label because nothing looked more pressable. The design of the pump failed to distinguish the data from the metadata.

The *Parade* advertisement, Wal-Mart, Shakespeare, and the gas pump are exceptions to the grace with which we usually navigate the implicit and the explicit in the first two orders of order. Human consciousness is built out of our ability to focus and be implicitly aware of our context simultaneously. With a flick of focus, the implicit becomes explicit, with its own implicit context. If we confuse the two, we may end up wearing a hole in the START label, but that rarely happens. We are masters of the dance of the implicit and explicit. But if you want your computer to do something, you can't hint, you can't type in an ironic tone of voice, and you can't gesture (although an application called "Bumptunes" lets you physically nudge your Apple PowerBook if you want it to go on to the next song). Computers deal only with what they've been told, not with what's been left unsaid. And that is causing a disruption of the delicate ecology of the implicit and explicit.

MAPPING THE IMPLICIT

To see what's underneath the play of the implicit and explicit, try to fill out a profile at Friendster, one of the first sites that let people create and expand their social networks online. The profile is typical, asking members to list their "hobbies and interests" so they can be matched with others. On my list are:

politics
Internet
reading

Then I ran out of interests.

I put "reading" on the list in part because it looks good. But that doesn't tell you much about me other than that I'm the type of person who puts "reading" on a list of interests. I didn't put "jazz" on the list because I don't know as much about jazz as someone who puts jazz on his list should. I didn't put my children's names on the list because that wouldn't mean anything to anybody else. I didn't specify that my interest in politics is especially in "U.S. politics" because that might make people think I am not an American. I didn't declare my unabashedly (and disappointingly predictable) politics because seeing "liberal" on the list would make it sound as if I only want to talk with other liberals. I enjoy movies, but I left that off the list because I am not more interested in them than anyone else is. I didn't put "television" on the list for the same reason. Besides—let's be honest—it wouldn't make me look good.

My interests aren't as pathetically narrow as my list would lead you to believe. It turns out I'm interested in the daily life of an eighteen-year-old Indian boy who sells paperbacks at a traffic interchange in Delhi, although I didn't know that until I read a well-written story about him in today's newspaper. I did know I was interested in the MIT Media Lab even before reading an article on its leader, Frank Moss, but it would be just weird if I were to put "Frank Moss" on my "hobbies and interests" list at Friendster. And I could not have known I was interested in the life story of four brothers and their drunken father in nineteenth-century Russia until I read *The Brothers Karamazov.* The novel created my interest.

My list of interests at Friendster isn't really a list of my interests. It's a complex social artifact that results from my goals, self-image, and anticipations of how other people will interpret my list. A frank discussion of how a person constructed her list would tell us more about that person than the list itself does. Just ask anyone who has struggled over what information to include in an online dating profile—and the businesses that have cropped up to craft profiles for lonely hearts.

Making explicit is *not* like moving something from the dark into the light. When sites like Friendster ask you to check a box to indicate if someone is your friend, that's often a decision, not a report on your inner social life. And because the other person will learn of your choice, it can hurt someone's feelings or give false hope. Making something explicit is often a social act with consequences.

Besides, friendship isn't that binary. I have no hesitation in listing my pal down the street as a friend, but that's not exactly how I'd describe the former boss with whom I had a good but not very warm relationship, the doctor with whom I chat but never see outside of his office, or the person with whom I've been exchanging intermittent emails about politics for the past ten years. If a site asked me the true-false question of whether they're my friends, I would probably say yes because they're not not my friends, but I would want to put an asterisk next to each answer. There's just so much more to say. In fact, there's almost always more to say than we can say explicitly.

The Friendster experience encapsulates much of the problem: Making complex, meaningful phenomena explicit can leave us rudderless, force us to oversimplify, and result in statements that are incomplete and misleading. We succeed at making things explicit by getting the balance with the implicit right. We do this when we give a hint while playing Twenty Questions, and we do it with far greater sophistication in the highly evolved technology that is an everyday map.

Steven Wright, the existential comedian, makes a joke about having a map at a 1:1 scale. The joke is funny—maybe more hmm-funny than haha-funny—because it makes clear that maps are useful only because of what they leave out. Howard Veregin, director of geographic information systems at Rand McNally, makes the same point just as directly: Maps "lie on purpose in order to tell the truth." He explains, "Even if a map were on a scale of one to one, there's so much more out there in the world that you can't map or is too dynamic or is too irrelevant. Sounds. Trash containers. Or the location of all the automobiles in Chicago at a specific instant. It doesn't capture all the richness of the world and it isn't supposed to." When Rand McNally makes a map for truckers, it notes highway weigh stations but leaves out national landmarks. Automobile road maps often show hospitals but not bowling alleys. Nor do they change the thickness of a line denoting a four-lane highway when the road narrows for half a mile, especially if a lane is temporarily closed for repairs.

Electronic maps are changing our expectations. ESRI is a half-billion-dollar company built on enabling organizations to create multiple overlays of electronic maps so a user can click a button and see, say, all the ranger stations in the forest, all the patches of deciduous trees, all the underground water reserves, and all the roads big enough to accommodate a truck. The maps shown on the current generation of GPS car navigation systems do something similar: As you slow down or as you near your destination, the system zooms in closer, revealing minor streets and their names. Going digital has enabled maps to show and hide information far more flexibly.

But there's still a world of information that doesn't make it onto maps. Your car's navigation system may plot where every gas station and every McDonald's is, but it doesn't show a Melville scholar the house where he wrote *Moby-Dick* and it doesn't circle the diner your friends were raving about on their blogs. Your company may have built useful overlays that show the geographic concentration of its suppliers, but it doesn't circle the ones who are pains in the butt. Electronic maps show only what their creators have programmed them to show, anticipating but not responding to users' purposes.

In 2005, Google Maps opened up some new possibilities. The maps displayed at the Google Maps home page make the same sort of decisions about what to show and what to hide as every other mapping site does, but Google did something plucky: It made it easy for other programs to incorporate Google maps into their own offerings, setting off an explosion of map-enabled applications. Some were straightforward: Hotel home pages that now can easily show the location of local services, travel sites that can automatically include a map of every destination they mention, businesses that show the distribution of their markets. But the ease of adding data and features to Google Maps spurred quick-witted developers to innovate. A couple of sites pull apartment listings out of CraigsList, the popular online classified ads site, and plot them on Google Maps. MapGasPrices.com shows all the gas stations near your home and how much they're charging per gallon. GMiF (Google Maps in Flickr) lets users populate a Google Map with the photos they took during their travels. Quickmaps.com lets users sketch routes as easily as drawing with a crayon, adding annotations to point out what they think is of interest.

A human rights activist posted a map of the prisons in Tunisia; click on one and it launches a video documenting their squalor.

Even as users become mapmakers plotting all the world's miscellaneous information, maps still work because what gets included and excluded is driven by a purpose. As Veregin says about Rand McNally, "We're interested in selling maps. Far and away what sells content is how we're targeting maps for specific audiences." When the purpose is not as clear, it's harder to get right what should be said and what should not be. The line between the implicit and the explicit isn't drawn by the intellect. It's drawn by purpose and thus by what *matters* to us.

WHAT IMPLICITLY MATTERS

The implicit can betray our real interests and way of thinking. For example, we know that Thomas Jefferson bought a copy of the Koran (known to him as "Alcoran") in 1765, when he was training as a lawyer. Even given the dominance of Christianity in colonial America, this is only slightly surprising, for Jefferson was famously curious about the world. He may have been led to the Koran by a standard law book written by the wonderfully named Freiherr Samuel von Pufendorf. But Kevin J. Hayes, a professor at the University of Central Oklahoma, noticed something about the catalog of literary holdings Jefferson compiled in 1783. Jefferson lists the Koran under the heading "Religion," but he seems to get the order of the list wrong. His religion category starts with works on pagan oracles, then lists the Greeks and Romans, then the Koran, and then moves through Judaism to Christianity. Since the Koran dates to the seventh century C.E., it should have come after Judaism and Christianity. Jefferson knew this; it was not a mistake, and we know that Jefferson took pains to categorize his books carefully.

Hayes argues that the clue is in a marginal note in Jefferson's copy of Gibbon's *History of the Decline and Fall of the Roman Empire.* Where Gibbon discusses the conversion of a Christian cathedral into a mosque, Jefferson wrote in a few lines from a poem about "a building returning to a state of nature," indicating that Jefferson considered the development of Islam to be a type of decay, says Hayes. (Before condemning Jefferson as a mere bigot, consider that not only did he try to teach himself Arabic, he drafted a bill in the 1770s that would have added "Oriental languages" to the curriculum at William and Mary College.)

Hayes's conclusion that "On [Jefferson's] library shelves and in his mind [the Koran] remained at a halfway point between paganism and Christianity" is all the stronger because it's based on implicit evidence. If Jefferson had explicitly declared that all religions were equal not only before the law but in their truth, the implicit lesson of his classification of the Koran would have belied that statement. The implicit often tells us more and is more credible than the explicit.

In Nick Hornsby's novel *High Fidelity* (and in the movie made from it), a mix tape is an expression of the mixer's feelings, longings, and personality—all without anyone filling in an explicit profile form. While the protagonist of *High Fidelity,* Rob, details some of his reasoning behind his playlists, he isn't able to explain it to the tape's recipient. Even if he could, a statement of that reasoning wouldn't have the same effect as listening to the songs would. The explicit often diminishes the implicit.

In the third-order world of digital music, the playlist is the descendant of the mix tape—except that a playlist is actually all metadata and no content, pointing to songs stored elsewhere. At the iTunes Store, users had created and shared more than 300,000 iMix playlists by February 2006, too many for people to sort through without the addition of some explicit metadata. So Apple allowed users to tag their playlists. You can find songs tagged "lonely," "Nascar," "breast cancer," and, of course, "love." A playlist is "an important means of self-expression," says Rebecca Tushnet, a professor at Georgetown University Law Center. "The motivation is an urge to say, 'This is who I am, and you can find out who I am by knowing what I love.' " Playlists aren't just a new unit of grouped music, they are deep expressions of self accomplished entirely through metadata.

Here's the odd thing about the implicit. Project teams often kick off with a round of explicit introductions—"I'm Carla from Quality Assurance. I'll be making sure what we build meets company standards"—but only once the members know more about one another than they can say has the group become truly a team. Likewise, good salespeople always know more implicitly about their clients than they can say explicitly—and certainly more than there's room for in the corporate Customer Relationship Management system. We can't say everything we know; that was the founding insight of Michael Polanyi, the Hungarian scientist and philosopher who coined the phrase "tacit knowledge" and who was an inspiration behind the knowledge-management systems adopted by businesses in the late 1990s. KM systems founder when they require employees to make explicit everything they know about a business, as if everyone is a natural writer or teacher—or as if everyone is a database that can generate reports at the push of a button. KM systems have done best when they've worked quietly, gathering the knowledge generated implicitly in the course of work, organizing emails into webs of information, inducing who is an expert based on seeing who's responding most frequently on in-house chats, and building libraries out of the links people send one another.

This goes beyond KM systems. If you were to ask me about my children, I would tell you some things. If we're talking about how our local public schools have been damaged by our culture's current obsession with standardized testing, I might tell you how one of our children does well on tests, another is a dyslexic whose brilliance is hidden by them, and a third is too free spirited to sit still for them. But I'll never be able to tell you everything about my children. If you want to stymie me, ask me directly to describe my children without giving the conversation a spin down some topical byway. I'll flounder just as I do when Friendster asks me to name my interests. In fact, if I could tell you everything I know about my children it would be a sure sign that our relationship is superficial. What I know of my children is too long and deep to be exhausted in words, too twisty, entangled, and intertwingled to be made completely explicit.

Dostoyevsky, though, does manage to tell us everything—or at least more than you or I could manage—about the four Karamazov brothers. We come out of the book feeling that we know them. Yet the miracle of art is that when you turn the conversation away from my children and ask me about the brothers in that Russian book I've been reading, I am equally unable to say everything I know about them. Somehow, through a series of explicit statements, Dostoyevsky manages to create an understanding of the brothers so rich and tangly that it defeats articulation. With lesser creations by lesser artists we can say everything we know: "He's the comic foil. She's a prostitute with a heart of gold"—and Carla is the quality-assurance person on the project. Try doing that for Ivan Karamazov, King Lear, or Carmela Soprano.

If the implicit is hard to talk about, the play between the implicit and the explicit is especially difficult to express, for in talking about something, we're making it explicit. Like the most sophisticated car navigation system ever invented, we're constantly making decisions about how much to say out loud and how much to leave unsaid—hiding and showing, giving details or leaving to the imagination, explaining or assuming it's been understood. Trees of knowledge work this way: If we know that this is a cat, by walking up the tree we can make explicit what is implicit in the organization of the tree—the cat is a mammal, an animal, a living thing, and a physical thing made of atoms. But when I see our cat sitting on a mat (which, by the way, is the standard philosophical example of a true statement), I implicitly know far more than that: The cat may soon get up from the mat, it's okay for the cat to be on the mat because the mat has already been chewed up, the cat likes the mat because it's soft and warm, the cat's resting on the mat is a sign of placidity in the household. The relationships touching the cat reach further and further, beyond any pragmatic possibility of being exhaustively listed: The cat came from a friend who moved to Hong Kong, the cat no longer chases birds, the cat was a consolation pet for our children after a hamster died, I have a good friend who is mildly allergic to cats, my parents wouldn't get me a cat when I was a child, *Catwoman* was a dumb movie. Each of those relationships touches an indefinite eddy of others in the Heraclitean swirl. Only one set of those relationships comes from the cat's explicit place in the taxonomy of animals, and in this case, the taxonomic relationship isn't particularly relevant. The relationships are unspoken, but under some circumstances they might be uttered, brought to light, made explicit, matter. Until then, they are inclusive, their moment of categorization postponed. The unspoken—the implicit—is *potential* to be mined, clustered, sorted, and mixed. It is the miscellaneous source of what we know and say.

MINING THE CLOUDS

If you want to know what Joshua Schachter is interested in, visit http://www.delicious.com/Joshua, his Delicious page. In the right-hand column you'll see all the tags he's used for the 8,505 bookmarks he's accumulated. With so many bookmarks and so many tags, Schachter has clustered his tags into categories he's created for himself, from "academic" to "time," although by far the largest category is "unbundled tags," the miscellaneous ones he hasn't put into a category. If you click on the "view as cloud" link, the list rearranges itself into an alphabetized paragraph, with the font size of each tag indicating in relative terms how many times the tag has been used. You can see at a glance that Schachter is more interested in "food" than in "coffee," in "humor" than in "forensics," in "art" than in "fiction," in "nyc" than in "sf." A tag cloud can read like a long haiku. Like a playlist or a mix tape, the truth is often hidden between what's explicit. We can go right or wrong in our sizing up of the person behind the cloud, but we are very likely to *go* because tag clouds visually express a person's interests, compiled from data the person communicated unintentionally. They are more likely to give an honest picture of a person than is a profile page or a job hunter's bio.

We can, of course, go wrong. People use Delicious for some interests and not others. And because Delicious pages are public, people may avoid listing some pages; I was too embarrassed to tag pages when I was researching large-screen televisions. Explicitly constructed profiles are at least as unreliable, though. We construct profiles based on who we think the audience is, what we're trying to accomplish—get a date or get a job—and, of course, what the profile itself asks us about. As Internet sociologist danah boyd says, "Most users fear the presence of two people on Friendster: boss and mother." There's often an element of self-delusion as well: Writer Sam Anderson maintains that the bottom of your queue of movies to rent from Netflix.com is "the person you want to be—*Eraserhead,* the eight-hour BBC *Bleak House,* the complete Werner Herzog—while the top is the person you actually are: *Wedding Crashers, Scary Movie 4, The Bridges of Madison County.*" Even that common mix of highbrow intentions and low-brow plans tells us something.

This opens up doors to marketers that we may not want opened. In the second-order world, direct mail—what those of us on the receiving end call "junk mail"—is often considered successful if 2 percent of the recipients act on the offer. Getting response rates even that high usually requires buying mailing lists carefully sorted by zip code and the recipients' history of purchases. In the third order, the amount of implicit information people generate about themselves is staggering. As AOL customers learned to their dismay in August 2006, once we know that a user AOL identified only as 545605 searched for "shore park margate nj," "frank williams md," "ceramic ashtrays," "transfer money to china," and "capital gains on sale of house," we're only a couple of guesses from knowing too much.

By pulling together implicit data from multiple sources, marketers can avoid being fooled by our lopsided self-presentations on any one site. But by mining the data, correlating it, even making guesses, marketers can know far more about customers than customers want; customers' leaves of information have been raked together without their explicit permission. How much of the implicit digital metadata people inevitably create should an organization track, keep, and use?

Although the ethos is changing rapidly, three norms are emerging about what's okay in the digital order and what isn't:

- Users probably understand at this point that a site may be recording their "clickstream"—the links they've clicked on at a site and even how long they paused on a page. But if a user has not registered with the site and logged in during a particular visit, then a business should assume that the user doesn't want her clickstream to be associated with any other information about her.
- Unless the user has given explicit permission, her tracks through a site should never, ever be shared with any other organization. (Explicit permission does *not* mean that the user failed to uncheck a box with some small print next to it on a page full of legalities.)
- Since the site is using *implicit* data that was not created in order to be tracked—users don't hesitate for 42.6 seconds before pressing a link *in order* to send a message to the company—a business operating in good faith will not use that information *against* the customer.

Granted, this last norm is tricky to apply because there are times when it might be in the customer's interest to be interrupted with an offer, even though she might ultimately reject it, but there are also times when it's obviously intrusive and even creepy. Amazon uses implicit metadata to put in links to relevant books, making it a better place to browse. And it enables customers to edit the metadata, excluding purchases that don't reflect their interests. A site that used implicit metadata to spam or embarrass a customer clearly would have crossed the line—and wouldn't improve its business.

The line is blurry because we are in transition in our idea of privacy and we are still discovering ways to make sense of the implicit traces people leave behind. But there is a line, and businesses who want customers to come back will pay close attention to it.

WHAT ISN'T SAID

A jar of jam you find at a roadside stand in Vermont has a label that says "Strawberry" and perhaps the year in which it was made. The photo of that same jar posted at Flickr might have tags that say "strawberry," "jam," "preserves," "cooking," "do it yourself," "Vermont roadside stand," and "gift." Yet the real jar tells us much more than Flickr tags ever could. The difference between the tags and the labeled jar is the world. The real jar is encountered in a real place in the real world. We know by looking at it that it's jam, that it's homemade, that it's for sale, that it contains sugar and strawberries, that you open it by twisting the metal top, that it has calories, that it goes well on toast. Tags capture only a few bits of that because tags by themselves have no context. Until we look at the picture, we don't even know if the tag "jam" refers to preserves, jazz, or traffic.

Therein lies a paradox of the digital order. As we pull the leaves from the trees and make a pile of the miscellaneous, we free the leaves from their implicit context. Compared to trees, piles of leaves are denuded of meaning. Tag a photo "robin," and the third order won't automatically know where it sits in the tree of species. The miscellaneous, with its tags and links, threatens to remove ideas from their context, diminishing their meaning and utility. Rather than knowing that a robin is a vertebrate, an animal, a living thing, and made of atoms, we are left knowing only what it's been tagged . . . and a tag may be as inscrutable as the photo of the front of a refrigerator that someone tagged "Capri."

But tagging is too young to be predictable. We don't even yet know if people will tag for themselves to refind pages or to help others find pages. Information architects—the professionals who design the organization of and human interface with information—have debated this ever since Thomas Vander Wal coined the term *folksonomy* in 2005 to mean an ordered set of categories (or "taxonomy") that emerges from how people tag items. Both schools of thought have supporters. Because Joshua Schachter thinks of his Delicious site primarily as an "amplification system for memory," he encourages people to use the tags that are meaningful to them, whether or not they are meaningful to others. For example, if San Francisco is your home and you find a Web page that lists local arts events that you tag "arts," you're unlikely to add "SF" as a tag because you take that for granted, which means your tags won't help tourists find the page. On the other hand, if you're a researcher tracking information about DNA, you may decide to tag a page about a newly decoded genetic sequence as "DNA," even though that's too broad to be useful to you, because you want to contribute to the stream of tags coming from—and for—the global community of geneticists.

The resolution of this dialectic between tagging for private use and for public good may come from the increasing power of computers to reconstruct the implicit on the basis of the explicit. Not everyone is hopeful. Information architect Peter Morville, author of the insightful overview of the state of the art *Ambient Findability,* says that the inability of folksonomies "to handle equivalence, hierarchy, and other semantic relationships cause[s] them to fail miserably at any significant scale." Yet there's reason to think that at sufficient scale—when we have an overwhelming number of leaves and a mind-boggling number of tags—our narrow-brained, literal-minded computers will be so good at rebuilding the context that the arts calendar not tagged "SF" could show up in response to someone's request to see what's going on in San Francisco. Four basic techniques are emerging.

First, someone else might have tagged the arts page as "sf." With enough people tagging, a computing system doesn't have to rely on the explicit tags created by any single person.

Second, the computer can learn from the sets of tags people apply to pages. If enough people tag pages as both "sf" and "golden gate," a computer can surmise there's a probabilistic relationship between those tags. It could also notice that many pages tagged "sf" are also tagged "california," and thus there's a likely relationship between "golden gate" and "california." It might also notice that the "california" tag is often used in conjunction with "san jose" and "los angeles" tags, and thus come to the tentative conclusion that "california" is the root of a tree that has "san francisco," "san jose," and "los angeles" among its branches. If the correlations are strong enough, when someone asks to see all the pages tagged "sf," the computer might also suggest pages that don't use precisely that tag, including perhaps the arts page. Flickr, Delicious, and Technorati (a blog index) use webs of inference of this sort to guide users to the pages and photos they're looking for.

Third, our computers may start to learn more about who we are, where we live, and whom we know. Intersect a tagging system with an online social network, and much of the context that tags ignore can be brought back in.

Fourth, there may well be clues within the arts page itself. It may say "Arts Council of San Francisco." Even if it doesn't, sophisticated software is able to figure out the city from the street names and the venues it lists. This can be done by seeing how street names cluster across millions and even billions of documents. It can also be done more systematically by building a gazetteer of place-names. Of course, it's not as easy as looking words up in a list of places. Most references to Pierre are probably not talking about the capital of South Dakota, and references to a small stone are not talking about the capital of Arkansas . . . not to mention that if a document talks about an event at the Polish parliament, the event happened in Poland, but if the event was held in a Polish embassy, it did *not* happen in Poland. Figuring out the *where* documents are talking about is a matter of gauging probabilities, but it can be done.

Of course, the computer could go wrong. If, for example, you take a photo of your growing son, Ben, in London, Ontario, and label it. "Big Ben," the photo may end up in a list of snapshots of London, England. If, however, the application had access to your online calendar, it could notice that you were in London, Ontario, on the day the photo was taken. It might also know that it was Ben's birthday. All this information could be used—with your permission—to disambiguate simple tags and to connect disparate threads.

Clustering has "pretty cool results," says Flickr's cofounder, Stewart Butterfield. He points to the system's automatic assembling of photos of noses into separate piles of dog noses and cat noses based purely on the "distribution and co-incidence of tags," not on software that can tell by the shapes whether it's a dog's nose or a cat's nose. Such techniques are good enough if you're looking for photos of dog noses or of San Francisco before you take a vacation there, but they are not yet sufficient if you are doctor who can't afford to miss a notice about a medicine's deleterious effect on the liver.

Flickr has another problem caused by its success. Because it has hundreds of millions of photos, and almost a million being uploaded every day, most are not all that interesting to the broad community of users. It's in Flickr's interest to feature photos on its home page that will be more arresting to the mass market of Flickr users. Implicit metadata turns out to be the key. Flickr watches the number of times a photo has been added to another user's list of favorites, the number of comments left on it by other users, the relationship of the commenter to the person who uploaded the photo ("A comment from your mom counts less than one from a stranger," says Butterfield), the number of tags, the number of times a photo has been viewed, and "a few dozen other things." Although at Flickr users explicitly designate other users as friends or as family, none of these factors can be easily manipulated by users to intentionally raise a photo's "interestingness" score; all are implicit in their behavior. But that is enough to ensure that Flickr almost always has arresting photos displayed on its home page.

So Peter Morville may have it backward: Tags may become more useful, meaningful, relevant, and clearer the more there are. If that is the case, the blind reasoning power of computers is only part of the explanation. Algorithms can find these relationships of meaning only because, just as all the items in our drawer of kitchen miscellany share the fact they are related to food, the items in the global miscellaneous drawer share a vast set of similarities in what we humans care about and how we talk about what we care about. Computers can cluster tags only because human interests and expressions cluster.

THE SPAN OF MEANING

Reading what's said to get to what isn't seems to be something we humans just do. A dog may know exactly what it means when it barks—"Get away from me!" or "I need to pee!"—but we humans rarely know exactly what we mean when we speak. If we were required to truly explain the full context for a remark as simple as "Wait a minute," we'd give up before we got to the division of time into uniform units and the concept of interrupting one intentional process with another that is more urgent. Each word echoes through the entire canon of language and social context. Without those echoes, we wouldn't know what "Wait a minute" means. What we can't and don't speak provides the meaning of what we do speak.

It is inevitable that we tend to focus on what is said and not on the unsaid that enables it, since as soon as we pay attention to the implicit, it becomes explicit. But we nevertheless have a vocabulary for talking about it. We call the implicit the *context* or *background*. We talk about our *assumptions* and sometimes even our *biases*. But there is another word for it, less obviously relevant: *meaning*.

Meaning's own meanings span a range unique in our language. On the one end, a meaning is a simple definition one can look up in a dictionary. At the other end, meaning is the broadest term for what gives value to our lives. There is a reason to think of it as the implicit and unspoken.

The German philosopher Martin Heidegger has never been accused of being easy to understand, but what he says about meaning makes sense. What does it mean to be a hammer? If I don't know that it's a tool for driving nails, I don't know what a hammer is. But if *all* I know is that it's for driving nails, then I still don't really know what a hammer is. I have to know at least that nails are metal spikes used for attaching wood. But my understanding of a hammer goes wider than that. I also know that wood is lumber made from trees, and that trees are plants that grow in forests, and that plants are rooted in the earth and grow toward the sun. To understand a hammer—not in some abstract way but as we grasp the hammer literally and figuratively—is also to understand that trees become lumber because we have an economy that pays people to do that work. And to grasp a hammer means to understand that humans have purposes because we have needs because we're not gods. It's also to remember the sound a hammer makes as it hits a nail, and perhaps the smell the board releases in response. All of this, says Heidegger, is part of our understanding of a hammer. The meaning of a particular thing is enabled by the web of implicit meanings we call the world.

Heidegger's is both a familiar and an unfamiliar view of meaning. It's familiar because when we talk about the meaning of an event, we are usually referring to how it fits into a broader context. It's unfamiliar because we have too often thought of words as sounds or scribbles that have meaning in the sense of a definition we can look up in a dictionary. But we look up words only when we're genuinely stumped. Almost always when we hear a word, we don't translate it into its definition—we don't turn every "hello" into "*n.*, a common greeting"—but instead hear its overtones, resonances, intentions, and connections. All that is *implicit,* however, so it seems odd and overblown when Heidegger connects a hammer to the sun and the economy. Nevertheless, I think his point is right: That implicit web of relationships gives the things of our world their meaning.

We humans have a history of extending thought into the world, as the philosopher Andy Clark points out in his book *Being There*. Calculators externalize arithmetic. Books externalize memory. As Marshall McLuhan taught us, sometimes the externalization not only extends but changes that which is externalized, the way books divided knowledge into discrete topics connected by cross-references to other books that may be aisles, floors, or continents apart. Books became *containers* of knowledge, as did experts.

Databases similarly externalized and transformed factual memory. The owner of a small store before the computer era kept a logbook of inventory. When she moved that information into a database, it got standardized in a way it may not have been before, the same columns defining every product entry. It also became possible for her to see relationships that the paper log obscured. Two facts separated by many pages in the logbook—there's a run on milk before every three-day weekend, beer sales have been trending down while white wines have been trending up—can leap out into obviousness once the owner has the database generate the right report. The physical nature of paper no longer gets in the way of understanding, although the type of understanding databases afford is limited by the nature of their contents: disconnected facts expressed precisely. The store owner can use the database to "what if" inventory levels but not to figure out if her morning customers didn't buy as many newspapers—or bought more coffee—because they were crankier than usual after hearing the local TV news.

Now, in the third order we are externalizing *meaning*. We can miss this when we refer to the digitizing and connecting of information as an "information highway" or as a vast library. Something more important is going on. In the third order, the content and the metadata are all digital. This enables us to bring any set of content next to any other, whether through relationships intended by the authors, crafted by the readers, promoted by the companies, or created by the customers. That makes the digital miscellany fundamentally different from previous miscellanies. The value of the potential, implicit ways of ordering the digital miscellany dwarfs the value of any particular actualization, whether it's how a researcher finds her way through ideas and facts to come up with a cure for a disease, how a citizen navigates through the laws and policies of her government, or how a customer leaps through a company's offerings to buy precisely the item that pleases her most.

We are building this connected miscellany link by link and tag by tag. Its value is in the implicit relationships that turn it into an *infrastructure of meaning*. From it we certainly can and do mine knowledge, treating it as the world's biggest (and sloppiest) database. But that's just the beginning. We populate databases by coming up with a standard set of columns and then stripping out everything that doesn't fit into them. To the digital miscellany, we're adding everything we can imagine, from a report on market preferences in Hong Kong to an MP3 of the band that played at the opening of the new

Hong Kong branch. And we're connecting leaves not merely through the statistical analyses our databases are whizzes at but through links, tags, playlists, and all the human ways we can imagine and invent, adding back the open-ended context that databases strip out. So, we can not only spot the statistical trend that shows that sales are declining in cities with populations over fifty thousand, we can read the blogs of customers who tell us why. The scientists in the R & D department are not only finding information by searching for it at Google, they're tagging it and creating streams—of information, explanation, speculation—shared every day with other researchers across multiple departments and perhaps even multiple companies. This infrastructure of meaning is always present and available, so that we can contextualize the information we find and the ideas we encounter. It is business's new greatest resource. Because it's shared by all—customers, partners, and competitors—the businesses that succeed will be the ones that embrace it most thoroughly and most intelligently.

The digital web of meaning has value to businesses only because it is about more than business. We are creating it not just as employees and customers but as citizens, parents, lovers, artists . . . all of what we are. That's what accounts for its richness and its fullness. Every phenomenon of meaning will emerge from the miscellaneous, from limericks to marketplaces, from new products to poetry to peace.

MESSINESS AS A VIRTUE

Since 1947, on a designated day in March eager brides-to-be have lined up outside Filene's Basement, Boston's foremost bargain clothing store. In a flurry of activity inevitably described by the local television news broadcasters as a "feeding frenzy," the brides rake through a couple thousand wedding dresses, marked down to as low as $249 from prices as high as $9,000. Women crowd the aisles, trying on gowns and bartering with one another. In minutes, the racks are bare. Gowns and street clothes are strewn about the tables and floor as if the proverbial hand grenade had gone off.

That's what the first order of mess looks like.

Medical records are a mess of the second order. Each hospital and clinic has had its own way of encoding patient information in its computer systems. Some might label a first name as "First_Name" and others might call it "Name01." Some might record an email address and others might not. Hospital records are so out of alignment with one another that doctors still fax paper to their colleagues instead of exchanging electronic information. Worse, as a patient moves through the system, there's no automatic way to link her records at her previous clinic to those at her new clinic. The second order of mess is, by its nature, a mare's nest of metadata.

As we straighten out first- and second-order messes, we feel better because we're restoring situations to the way they ought to be. As the moral term "ought" suggests, restoring order touches a sense of propriety that is deeper—or perhaps just older—than our need to tidy up. In restoring order we are making the world habitable, fit for humans. Messiness is a disruption. Orderliness is the way things are supposed to be. It is the Eleventh Commandment, the one that caused the other ten to arrange themselves in neat lines on two symmetric halves of the tablet.

There are practical benefits to being well-ordered, of course. A library card catalog dumped on the floor is of no value. But arrange the cards in neat alphabetical order by author, title, and subject, and now you can find the books you're looking for and browse by topic to find the books you didn't know existed. Sometimes you might have to disrupt alphabetical order to make books findable, as Thomas Hyde, the librarian at the Bodleian Library at Oxford, did in 1674 when he boldly lumped "Shakespeare" with "Shakspeare," "Shakspere," "Shaxberd," and the other dozen or so variants. Hyde messed up alphabetical order to clean up a spelling mess.

Organizing things neatly in the first two orders requires us to make those sorts of decisions. In the first order, we have to pick one way of arranging the objects, and that one way will not suit every user and every need. The second order allows us to add a few alternate ways of organizing the information (by subject and title as well as by author, for example) but it is feasible only if we strip out most of the information in the first order: A card catalog reduces a book to what fits on a three-by-five card.

The third order, on the other hand, is a mess from the git-go.

Imagine digital cameras have not yet been invented and that you've just picked up a roll of film from the local developer. You shuffle through the twenty-four photos quickly while sitting in your car, throwing out the really awful photos—the ones with your thumb over the lens and Aunt Sally with her lipstick smeared. At home you transfer the rest to the shoe box that serves as your holding pen. Eventually—for some of us, the eventual takes decades—you get around to selecting the best of them, and put them into your photo album. You put those that don't make it into the album back into the shoe box, where they'll rest comfortably, undisturbed in rough chronological order, until your heirs decide they need some closet space.

That miscellaneous pile of photos in your shoe box contains the photographs you think *might* have meaning to you or your family at some point. They are a *potential* source of memories, if not for you, then perhaps for your children. Maybe your kids are in the background of that awful photo of Aunt Sally and it brings back a surprising moment from their childhood.

The first-order mess you're leaving your descendants drives down the value of the shoe box. The more photos you add, the less likely that you're going to be able to find a particular photo, and the bigger the hurdle to making the pile usable. If you were to impose a second order of order by drawing up a careful index of every photo and its contents, the chances of anyone realizing the potential in your collection would go up, but at the cost of too much effort on your part, since these are the photographs you didn't deem worthy of a place in an album. The potential of this first-order and second-order miscellany likely goes unrealized.

But if you're taking digital photos, you're building a third-order mess snap by snap. As you dump hundreds and then thousands of digital images onto your hard drive, the mess gets worse and worse—but the potential gets greater and more realizable as you add metadata to the photos so that they become ever-smarter leaves. Their potential will increase even as they become no less messy.

That fact ends the argument that Oscars and Felixes—*The Odd Couple*—have been having in one form or another ever since Adam thoughtlessly tossed his fig leaf onto the ground and stretched out to scratch himself in the Garden. We all alternate between making messes and neatening them up because in the first two orders, things have places, and as we use them, they get out of place. But there are no places for things in the third order. As we saw with the computers that house Wikipedia, the physical placement of the bits is of so little importance that even the people in charge usually have no idea where they are. This is a mess of a whole new type. In the first order, if Felix decides to arrange the kitchen utensils alphabetically but Oscar comes in and shoves them all into an empty beer carton, a comic struggle will ensue. On the other hand, if Felix decides to neaten up their *digital* stuff by assigning each one a special numeric code so he can list them up, down, and sideways, whereas Oscar prefers to tag the stuff with the type of cigar he was smoking when he first encountered each one, it doesn't affect the other roommate's order. We straighten up third-order messes by arranging their metadata, leaving the actual objects untouched. The same digital miscellaneous pile can be as orderly as an operating theater to Felix and as messy as a landfill to Oscar.

Even better, each and every way Oscar, Felix, Adam, Eve, and everyone else straightens up the pile adds value to the mess. You can see this at Flickr, the photo-sharing site. As you upload your photos, Flickr automatically captures metadata hidden in the digital photos themselves: Who uploaded them? When? With what type of camera? At what exposure and focal length? Was the flash used? What was the zoom? That information can be useful should you want to find all the indoor photos you took after six P.M. in March 2005. And that's before you've started adding tags, writing descriptions, and grouping photos. Then your social group adds their own tags, designates some as favorites, writes comments, and leaves implicit metadata—they printed one photo and viewed another each day for two months—that signals how they feel about your photos. Because Flickr knows this not just about any one photograph but for the hundreds of millions of photographs it carries and the more than 540 million tags, it has a vast array of interrelated metadata that enables it to compute the connections and intersections that mark the implicit, shifting social network hidden in the mess.

Flickr is a mess that gets richer in potential and more useful every day. If Flickr were, for example, to add a gazetteer so that it could show you photos marked "Broadway" when you search for photos of New York City, it would have even more metadata that could be intersected with the rest. If it added face-recognition software, yet more relationships would be available. The more metadata, the messier and richer the potential. Third-order messes reverse entropy, becoming more meaningful as they become messier, with more relationships built in.

Messiness has always been with us, of course. But our culture has not only struggled against it, it has measured its progress by how thoroughly it has tamed it. Everything has its place, we've been taught, and we master our world—we *know* it—by discerning and enforcing those places. The challenge to this regime of orderliness certainly does not arise only from the digitizing of information. As we will see, psychologists examining how humans actually categorize what we experience have discovered that our theories have been wrong in significant and measurable ways. But it is the digital order that enables us to make a far bigger—and far more useful—mess than ever before.

SCRIBBLING IN THE WHITE SPACE

In December 1953, Dwight D. Eisenhower addressed the United Nations General Assembly on the subject of nuclear bombs:

> It is not enough to take this weapon out of the hands of the soldiers. It must be put into the hands of those who will know how to strip its military casing and adapt it to the arts of peace.

He concluded his speech by calling on the United Nations to create an agency that would collect fissionable material from the atomic powers—the United States and the Soviet Union—in order to develop peaceful uses for nuclear technology.

But who would govern this new agency, which would be a third nuclear power, albeit one devoted to peace? Following President Eisenhower's wishes, the International Atomic Energy Agency, as it was named, to this day reports to the UN's Economic and Social Council. But its organization chart has a dotted line conspicuously running to the UN Security Council. This thin, broken thread was inserted because the USSR insisted that the IAEA report to a unit of the United Nations over which it had a veto. As is so often the case, behind the dotted line that mars a neat organizational scheme is a story of power and fear.

When Daniel C. McCallum drew the first organization chart of a modern corporation, in 1855, it had no dotted lines. McCallum was not the first to restructure a railroad. After a head-on collision had killed a conductor and an engineer, the Western Rail Road reorganized itself into a hierarchical management structure reminiscent of the U.S. military. Soon after, so did the Pennsylvania. In both cases, the reorganizations were led by West Point graduates. McCallum faced a different imperative. The Erie was in financial difficulties after consolidating with the New York Central. McCallum, an engineer who had risen to become the general superintendent of the Erie, recognized that railroads were too geographically spread out to be controlled entirely from the top. It would be like an emperor in Rome thinking he could micromanage the units fighting in India. The information just didn't flow quickly enough. So he divided the company into regional units, creating more managers and increasing their importance. McCallum stressed "that channels of authority and responsibility were also channels of communication," observes historian Alfred Chandler Jr. Indeed, McCallum's fifth principle of administration makes explicit that his organization chart—recognized as the first diagram of a modern corporate structure—intermingles the flow of information and authority:

> Such information, to be obtained through a system of daily reports and checks, that will not embarrass principal officers nor lessen their influence with their subordinates.

To accomplish this, reports went only to the author's immediate superior. His chart established not only an information flow but also formal information blockages.

McCallum's system worked. The railroad knew exactly where its trains were at any moment of the day, and reports could be studied over time to discover and remove inefficiencies in the system. For example, the railway was able to adjust prices to encourage traffic on underutilized portions of trips. His chart became famous. The *Atlantic Monthly* ran an article praising McCallum's ideas. The editor of the *American Railroad Journal* sold copies for a dollar each. The ability to represent a complex enterprise in a simple diagram made McCallum a star because his way of thinking about business went beyond the particulars of the hierarchy at the Erie railroad. McCallum bestowed the workings of business with the properties inherent in the diagram he drew:

1. McCallum's chart was *simple.* Everyone knew to whom they should give the detailed hourly, daily, and monthly reports the system required.
2. It was *uniform.* A line going into one box meant the same as a line going into another box.
3. Everyone in the organization had a *place.*
4. That place was a *system* in which the parts had well-defined relationships.
5. The chart made the system *explicit.*

Simple, uniform, comprehensive, orderly, explicit: That's what we mean by neatness. And its benefits are obvious. Whether we're talking about a corporate organization chart or a kitchen drawer, we can find existing items quickly, and we can easily assimilate new items into existing categories. This helps accomplish the biological aim of categorization—dealing rapidly with an ever-changing environment by assimilating the new to the already established. Beyond the effect on our fitness as a species, a neat environment gives us a sense of mastery. The compound we establish in the land we've cleared is neater than the wilderness around it. We make order; disorder happens when we lose control.

Neatness has also been a characteristic of our systems of knowledge. Linnaeus straightened up the messy house of creation so it fit into three double-paged spreads of boxes within boxes. Mendeleev played chemical solitaire until he found a way to lay out the elements in a grid. Dewey reduced the knowledge in the world's books to ten categories, each divided into ten and then into ten again. Knowledge has belonged to Felix, who not only puts things into their place but devises the simplest, most elegant system of places imaginable.

In such systems, exceptions are regrettable. In the case of the International Atomic Energy Agency, it took the strong-arm diplomacy of a nuclear-armed superpower to get a dotted line drawn to a division of the United Nations it could control. If there are too many dotted lines on a traditional organization chart, it's taken as a sign that the chart needs to be redrawn or the management team needs to be replaced. Dotted lines traditionally are a sign of failure.

That's changing. For Valdis Krebs, a pioneer of the new social cartography, a neat map hides more than it shows. "There's a lot of white space on an org chart," he says. "My group finds out what goes on in that white space, and we fill it up with colored lines." The colors represent the departments people are in. In one case, Krebs analyzed who was sending email to whom and discovered that among the people working on a large project that had fallen months behind, the colors were clustering too neatly, indicating that people weren't talking with their teammates in different departments. Worse, only a small handful of people had lots of lines going into them. It turned out that team members felt they had to route questions and ideas through their own department leaders rather than go directly to teammates from other departments. So, at Krebs's suggestion, the project managers moved people's desks to alter the existing relationships and create new ones. Soon the dynamics of the group changed. The department leaders who had had too many lines and too much email coming into them were able to get more of their own work done, and the project got back on track.

These new maps of social connection demonstrate the value of messiness. On a typical map, each person is represented as a small square. Lines representing some form of social interaction spider out from person to person. It's immediately apparent which people are hubs because they're the ones who look like the center of a spectacular bloom of fireworks. When IBM called in Krebs to figure out why a project wasn't going well, he did a quick survey to find out who talked with whom when there was a work-related problem, who worked together on a daily basis, even who were the usual sources of rumors.

The map of those human connections exposed the fruitful messiness behind the neat map of the command-and-control structure.

Frequently, Krebs's maps contradict the maps of formal authority. For example, in the late 1980s, Krebs created a social map for his employer, TRW. Looking at it, one of the senior managers told Krebs that there must be a mistake because the hotshot they had hired from the air force had only a few links coming into him, but another person—"Let's call her Mary," says Krebs—was at the center of the social ecosystem. At first the managers disputed the accuracy of his research. "I said we'd checked our data," he recalls. "There was a silence. Some people were staring at the ceiling and others at their shoes. Finally someone spoke up and said the reason the air force guy doesn't have as many links is that he's a jerk." While Mary didn't know as much as the jerk did, she welcomed people into her office. Because she knew and liked her coworkers, she didn't waste their time explaining what they already knew. And she understood how the place worked. "With Mary, you walked away smarter, with a list of questions and steps you needed to take," says Krebs. The org chart, with its single, simple line into and out of Mary's box, failed to express the value her complex social relationships brought to the company. On old-fashioned paper she looked less valuable than a hotshot jerk sitting alone in his office, because the standard org chart hid her key role in the messy social world where most of the real work of the office gets done.

So why do we have so much white space in the standard org chart? "Part of it is that we abhor complexity," replies Krebs. "We try to keep our lives as simple as possible. We think if we logically organize things, then everything will slot into those holes, and we'll be able to face any situation. That was maybe true in the 1800s, but now we're surprised all the time."

Krebs points to another crucial benefit of enabling and encouraging conversations across the formal lines of authority. "Network analysts have found that innovation happens at the intersections," he says. "Ron Burt of the University of Chicago studied Raytheon and found that those people positioned correctly in a network couldn't help but get more ideas than someone who wasn't positioned there." "Positioned correctly" means being at the intersection of ideas. "If you're in a busy intersection in a city, you're more likely to get splashed with water," Krebs explains. In a business, at those messy crossroads you're more likely to get splashed with ideas, even though those intersections usually don't even appear in the official organization chart. The messiness of a diagram of social interaction is often a measure of the level of innovation in a company.

Each company has one official org chart because the flow of authority needs to be simple and unambiguous for legal reasons, not just to create an efficient decision structure. The chart works in those capacities *because* it has so much white space. But in the emptiness that simplicity requires, how many different social networks actually exist? "How many do you want?" Krebs answers promptly. Do you want to map who works with whom, who talks with whom, who knows whom, who respects whom, who enjoys whom? Do you want to map the path of memos, emails, instant messages, phone calls, and hallway conversations? Do you want to map it all over time? Social networks are necessarily loose-edged and impossible to make fully explicit. If your aim is to come up with a map as simple, uniform, comprehensive, orderly, and explicit as McCallum's, you need to ask only one question: To whom do you report? But if that's all you see, your world is neat, incomplete, misleading, and boring.

Simplicity was the only reasonable strategy before we developed machines that can handle massive amounts of data and metadata. Smart businesses are no longer confined to knowing what can be written in two-dimensional lines on the flat surface of a sheet of paper.

ESCAPING DEFINITION

Aristotle would have been more at home with Krebs's messy maps than most Aristotelians would be. In his *Politics,* Aristotle tries to describe the best constitution without "assuming a standard of virtue which is above ordinary persons." He may have thought that rationality was of the human essence, but he knew that in fact we are twitchy mixtures of reason, emotion, and desire.

Aristotle held knowledge to a different standard. If you want to propose a political system, you have to deal with people as they are, but if you want to penetrate to the truth about the world around you, you can't settle for appearances and blurry lines. To know what a thing is, thought Aristotle, is to see what is essential about it (that humans are rational animals) and not be fooled by just what happens to be true about it (that humans have their navels on the front). The definitions of those essences determine which things are in a category and which are turned away. Here there is no messiness, only an order so precise and harmonious that it is beautiful.

Or so Aristotle and generations of thinkers assumed. So do we when we argue about, say, how to define race, knowledge management, or blogging. But suppose this sort of Aristotelian categorization-through-definition were shown to be an essentially artificial way of approaching the world. Suppose the neatness it strives for is impossible. Suppose messiness is not a flaw in our thinking but enables it.

In her office, lit only by the late-afternoon light slanting in through the window, Eleanor Rosch turned back my question about the overall significance of her work: "What do *you* think its significance is?" she asked. In a different tone of voice, from a person seated less squarely or dressed less practically, this might have been a request for praise. Instead, it seemed to be a way to get at why I had come, as well as a dodge by a person unwilling to speak as immodestly as my question proposed.

I paused, unprepared. "I think you unhorsed Aristotle."

This isn't a matter of pulling down a dusty equestrian statue. When I asked for an example of Aristotle's continuing influence, Rosch said, "For the past two and a half days, I was at a conference on the effect of the media on the Buddhist transmission into our culture. Attendees kept asking, 'Wouldn't it help if you first defined Buddhism?' By that they meant an Aristotelian definition. If that's what we need, then the conference couldn't have happened." She continued: "As far as I can see, there isn't a single course that could be taught at this or any other university"—Rosch is in the psychology department at the University of California, Berkeley—"if we had to start out by defining the subject matter. No one at the conference could define Buddhism, but no one had the least doubt about what the conference was about."

Rosch was on an unanticipated stop—New Guinea—in her life's itinerary when she had her fundamental insight. When she was seven, her family had moved from New York City to Tucson to accommodate her father's sinus problems, and then to San Fernando. At Reed College she worked on a joint major in philosophy and psychology, but when her psychology professors didn't like her honors thesis on Ludwig Wittgenstein, she was graduated as a philosophy major. While doing her graduate degree at Harvard's Department of Social Relations, an interdisciplinary unit that combined sociology, psychology, and anthropology, Rosch met and married an anthropologist. The newlyweds got funded to go New Guinea.

There Rosch studied how one of the local tribes, the Dani, categorized color. Color categorization was an interesting field for anthropologists and linguists because the 7.5 million colors humans can perceive form a continuum with seemingly no natural divisions. Yet pioneering work by Brent Berlin and Paul Kay in the late 1960s showed that across 110 different languages, there seem to be only eleven basic color categories. Cultures disagree about which of the colors count as basic—Russian has no single word for blue, French has no single word for brown, and the Dani, remarkably, have only two basic colors—but the sets of colors all seem to be drawn from within that group of eleven. Rosch showed the Dani a color swatch and thirty seconds later asked them to pick the color from a handful of others, and then repeated the process with a new color. She found that they identified the basic colors more accurately than the nonbasic colors. So do Americans and subjects from twenty-three different language backgrounds. It seems that although we disagree about how many basic colors there are, when we lump and split we identify some swatches as prototypical examples of colors and others as sort-of, kind-of, to-some-degree examples—this swatch is a "pure" red but that one is reddish.

This flies in the face of the Aristotelian idea. For Aristotle, a thing is a member of a category if it satisfies the definition. Thus, anything in a category is an equally good example of it. After all, it shares the *essence* of the category. But when it comes to color, it seems that we don't work that way. Tomato red is a great example of red, the sort of red you could point to if someone didn't know what "red" meant, but a maple leaf in autumn may be a dark red, an orangey red, or a brownish red—it's red, but not a good example of red.

Rosch wondered if this was true more generally. In the physical world, we have to lump and split in an Aristotelian fashion, storing our laundered sweatpants with our work pants or with our sportswear. Conceptually, however, we may categorize sweatpants as a not very good example of pants but a pretty good example of sportswear.

Rosch went back to the subject of her college honors thesis, Wittgenstein, for an alternative to Aristotle. Wittgenstein, one of our most imaginative philosophers, famously asked about the meaning of "game." Scrabble, poker, solitaire, football, bingo, and a child counting how many times she can catch a ball she's thrown into the air are all games, but they have no single feature in common, and thus no definition works perfectly to include everything we consider to be a game. Instead, Wittgenstein said, games have a family resemblance: Carl has the family chin, Carla has the family eyes, and Carlita has the family ears, but no family feature is present in all those who share the family resemblance. In the same way, some games have teams, some have winners, some have rules, but there is no single set of features they all have, and thus there's no Aristotelian definition of "game." Nevertheless, we all know what the word *game* means. So, apparently we can know what something means even if it can't be clearly defined and even if its boundaries cannot be sharply drawn. Rosch realized that concepts can be clear without having clear definitions if they're organized around undisputed examples, or prototypes, as she calls them. That's as radical a thought within cognitive psychology as Wittgenstein's family-resemblance theory was within philosophy.

Rosch began examining exactly how prototype concepts work. Why is it, she wondered, that if I ask you what you're sitting in, you'll say, "A chair," and not "Furniture." If I ask, "What are you driving?" you're likely to say, "A car," and not "A vehicle," although depending on the context you might also say, "A sports car." What makes something a *basic-level* concept such as "chair" and "car"?

Rosch hypothesized that since "the task of categorization systems is to provide maximum information with the least cognitive effort," basic-level objects (chair, car) should have "as many properties as possible predictable from knowing any one property." That way, by knowing that something is a member of a category, you would know a lot more about it, the type of efficiency that natural selection would seem to favor. On the other hand, she hypothesized that "superordinate categories" (furniture, vehicle) would "share only a few attributes among each other." She tested nine sets of categories—tree, bird, fish, fruit, musical instruments, tool, clothing, furniture, and vehicle—asking subjects to list all the attributes they could think of at three levels of abstraction (e.g., vehicle, car, sports car). Sure enough, they came up with very few attributes for the superordinates, but lots for the basic-level categories.

She did more research. Superimpose the outlines of examples of a basic-level concept, such as chairs. Then average them so you get a generalized chair shape. Then do the same for outlines of its superordinate, furniture, including chairs, beds, and couches. The outline of the basic-level chairs is much more recognizable than the outline of furniture in general. The basic-levels seem to be the core categories: The basic-level names are more frequently used than the superordinates, they are the basics of the folk taxonomies used by less developed cultures, and even two-year-olds sort well by basic-level categories, although they use different categories than adults do. (By the time children are four, they sort superordinates with 96 percent accuracy.) In fact, another experiment showed that people can describe physical interactions with basic-level objects better than with superordinates—for a chair, you'd make a sitting movement, but what movement would you make for "furniture"?—suggesting that basic-level objects are connected not merely to mental manipulations but to how our bodies operate in the world.

Yet concepts don't just have other concepts above and below them in the Aristotelian tree. They also have other concepts next to them: cars next to trucks next to bicycles, all hanging from the "vehicles" superordinate branch. And here's where our way of understanding the world gets really messy.

"Most, if not all, categories do not have clear-cut boundaries," Rosch declared in the face of the most basic assumption of cognitive psychology. In support, William Labov reported in 1973 on research that showed that where we draw the line between cups and bowls depends not just on their shape but whether we're imagining them filled with coffee or mashed potatoes. Later researchers have found the same sort of continuum in verbs such as *look, kill, speak,* and *walk,* and in abstract categories such as "tallness." But the lack of clear-cut definition does not mean that we are set adrift in a "blooming buzzing confusion," as psychologist William James described a baby's sense of the world. Show us a clear-cut example—a prototype—of a cup or a bowl and we feel no ambiguity at all. Likewise, in our culture we agree that a robin is a good example of a bird, but an ostrich, flamingo, or penguin is not. A kitchen chair is a good example of a chair, but a beanbag chair is a terrible example of one. A car or a truck is a good example of a vehicle, but a skateboard is not, and skates aren't vehicles at all, although it's hard to explain why not. The prototypes—the good examples—do the job of organizing our world that Aristotle thought required essences and definitions.

Of course, which items are the prototypes is culturally relative. Research shows that for Americans, the best examples of furniture are chairs and sofas, while Germans think beds and tables are the best examples of their word for furniture (*Möbel*). But that we think in terms of prototypes does not seem relative. Within any culture, what makes one type of bird—perhaps a robin—a prototype of birds and another—a penguin—not? Rosch hypothesized that prototypes have more features in common with other members of the group than nonprototypes do. Imagine a Wittgensteinian family resemblance. If Carlos is a prototypical family member, he's got more of the features that are distinctive of the family: the high forehead, the crooked smile, the crinkly dark eyes, the elfin ears. The family traits are present but less pronounced in cousin Carla, who's got the ears and the smile, but not the forehead or the eyes; cousin Carla has the family resemblance but isn't a prototype of it. Tests on four hundred students in introductory psychology classes proved Rosch's point. She found that the students were able to list more features for basic-level words than for superordinates; knowing that something is a bicycle brings more associated knowledge with it than knowing that something is a vehicle. The results also strongly supported Wittgenstein's family-resemblance idea: There was no set of attributes shared by all the members in a superordinate category such as "vehicles," even though in follow-up interviews students insisted that there must be.

Rosch's findings stand in stark contrast with the prevalent definitional view that thinks we start with criteria and then find some good examples. The prototype view thinks we start by having prototypes pointed out to us—"Oh, look, a birdie!"—and then cluster other things around them. The definitional view draws sharp lines. The prototype view works only because things can be sort of, kind of in a category, the way a skateboard is sort of a vehicle. Prototype theory relies on our implicit understanding and does not assume that we can even make that understanding explicit.

Sometimes we do have to draw sharp lines. A riding lawn mower, a motorized wheelchair, and a battery-powered skateboard may all sort of be vehicles and sort of not, but the Department of Motor Vehicles has to make a yes-or-no determination when you drive up to the local office and ask to register it. We can stipulate a definition that will work at least pretty well, even though it's arbitrary and artificial. But that's not what experience looks like. First comes a hands-on, body-and-soul roughhouse of organization built on multifaceted resemblances to clear examples. Lines come later, and only when we're forced to draw them. Where and how sharply they're drawn has everything to do with who is drawing them and why.

This means that a business that forces its products—or its

employees—into a predefined set of categories is performing an unnatural act. What the business insists is an orange ski hat that belongs in the sportswear section its customers may see as a marigold knit cap that belongs in the urban section. Keeping products miscellaneous allows customers to search for them more efficiently. Allowing customers to tag items lets the products be in multiple categories at once. Watching customers' browsing and buying patterns enables places like Amazon to pull together offerings across category lines. Thinking that people's skills are defined by the department they're in wastes their talent. (It also means that companies frequently start corporate blogs with the least interesting people—the marketers—as their initial bloggers.) As quickly as a business neatens up—so accounting systems can do their job, if for no other reason—it should scribble over the lines of division with lines of connection. Every line that's drawn ought to be systematically smudged. For the fact that we think in prototypes means that messiness isn't a flaw. It's a strength. We *can't* put everything away in its place because those places are just sort-of and kind-of where things belong. Everything belongs in more than one place, at least a little bit.

Eleanor Rosch's research shows that messiness begins within. It is, so to speak, of our essence, and to imagine thinking without mess is to imagine thinking the way computers think, which is to say, it is to imagine not thinking at all.

THE SEMANTIC MESS

There is no dorm room, divorce, or political scandal as messy as the World Wide Web. There's an excellent reason for this: Sir Tim Berners-Lee, the inventor of the World Wide Web, in his wisdom made sure that the Web is a permission-free zone. Anyone can post anything she wants, and anyone can link to anything else, all without alerting a central registry, without having to get approval, and without anyone saying exactly where to shelve the new material. So, the Web has grown without a plan, which is exactly why it has grown like crazy.

Now Sir Tim—the title sits awkwardly on this modest man—looks out at his creation and wishes it were neater. But the grand vision of his cleanup plan—the Semantic Web—invokes the Genie of Taxonomy, our old urge to build an organizational structure so big that everything fits into it and nothing is left out. As the Semantic Web encounters the deep mess of the World Wide Web, we will learn which techniques familiar from the first two orders can be reshaped for the third.

One can trace the roots of Berners-Lee's dissatisfaction with the Web back to a software program he wrote at the beginning of his career that eventually inspired the Web itself. Named, he says, after *Enquire Within upon Everything,* a "musty old book of Victorian advice I noticed as a child," Enquire doesn't just keep a list of the parts of a particular machine or the people working on a particular project. Rather, it tracks the context of relationships among the people, parts, and information so that users can know not only that something is, say, a handle, but that it is part of the cranking assembly, that it includes a particular ball-bearing assembly, that it turns clockwise, that it's made out of iron, and that it was produced by the Acme Crank company. To achieve this, Berners-Lee designed Enquire to include relationships such as "made by," "includes," "uses," "describes," "background," and "similar to."

Enquire may have been a useful tool, but it's not much like the Web it inspired. In fact, the example Berners-Lee gave when first writing up Enquire—the prototype, in Rosch's sense—is of a vacuum-control system, rather distant from the ethereal, loosely woven material of the Web. There wasn't anything in his explanation describing home pages, e-commerce, or weblogs, yet Berners-Lee says that Enquire "gave rise to the idea of the World Wide Web." He writes:

> Suppose all the information stored on computers everywhere were linked. . . . Suppose I could program my computer to create a space in which anything could be linked to anything. All the bits of information in every computer at CERN, and on the planet, would be available to me and to anyone else.

Links. Now we're in the Web's home territory.

Of course, there was nothing new about links themselves. In a 1945 article in the *Atlantic Monthly,* Vannevar Bush proposed building what he called a "memex," "a device in which an individual stores all his books, records, and communications." The memex would be embedded in the desk of the future, onto which would be projected images from microfilms of works as the user accessed them via a keyboard. "All this is conventional," Bush said, knowing that his readers would be dazzled. The real advance would come with "associative indexing": The user could pick any items from his microfilm index and associate them. This way the user would build up "trails" that would persist over the decades and could be built into a shared universal associative library. Like Enquire thirty-five years later, the memex was all about the links.

But memex links and Web links, unlike Enquire's, don't say anything more than that A *points at* B. Put simply, Enquire is about smart links—links with metadata—while the Web is about dumb links. So why did Berners-Lee dumb down Enquire when he came up with the World Wide Web? The answer is in the inspiration: *"Suppose all the information stored on computers everywhere were linked."* What would that take? You'd need a network that connects the computers. That would be the existing Internet. You'd need a way to put a link in a document that points at the Internet address of another document; for this, Berners-Lee invented HTML, with its ability to turn any string of text into a blue underlined link. You'd need software that could display the documents written in HTML; Berners-Lee wrote the first Web browser. And now you'd have everything in place for a worldwide web—except for the pages that are the Web's substance.

To get the world to start writing Web pages, you'd have to make it astoundingly easy to create and link them. And here Berners-Lee's genius really hit its stride. Give up on management, he decided. Let anyone link to anyone else without having to ask permission or to get it categorized. Just post it. Just link to it. Click. Done. Within five years the Web became the largest aggregation of human intellectual creation in the history of our species. To enable the Web to become worldwide, Berners-Lee had to make it far stupider than Enquire, his first labor of love.

Apparently, it's been bothering him ever since. The Semantic Web aims to make the Web smart through the power of metadata, although sometimes in ways that are reminiscent of second-order attempts to categorize the universe.

Berners-Lee begins his *Scientific American* article on the Semantic Web with a futuristic scenario in which Pete and Lucy's software agents negotiate complex arrangements for medical treatment for their mother. The right local doctors are found, appointments are made, and transportation is arranged, automagically, because all the relevant pages have been tagged with metadata. The software can find exactly which doctor is how far from home, what her specialization is, when her next open appointment is, and when the local buses run. This implies a massive agreement about exactly how the metadata should be expressed. Everything works perfectly if everyone just agrees on the terms and follows the rules.

It's the old dream of rationalism. Build a comprehensive system and drive the ambiguity out of it. Minimize the miscellaneous. Turn language into a machine and our machines will work wonders. People are too fallible. It is a dream that has failed over and over, not for a lack of trying. But Berners-Lee thinks this time he's cut the Gordian knot. Remove central control over the "definition of common concepts such as 'parent' or 'vehicle,'" he writes, because "central control is stifling, and increasing the size and scope of such a system rapidly becomes unmanageable."

Here the Semantic Web takes a giant conceptual step beyond Berners-Lee's first system. Enquire stipulated a set of acceptable relationships among the things it linked: *includes* and *describes* but not *owes money to* or *hates the smell of*. The Semantic Web proposes not a standard set of relationships but a standard way for people to describe whatever relationships are important to the topic. Called Resource Description Framework (RDF), this standard lets metadata be expressed in "triples," two terms connected by a third: Cars *are a type of* vehicle, Tricia *is a daughter of* Richard Nixon. RDF opens wide the limited set of relationships Berners-Lee hardwired into Enquire.

All those possible relationships, all those ways of expressing them! Why, one could capture the entire world! And that's just what some proponents of the Semantic Web are trying to do. A set of RDF triples that describes a particular domain—medicine, astronomy, cooking, stamp collecting, international finance—is called an ontology, and some are rather ambitious. For example, Legal-RDF has a "basic" vocabulary that contains more than fifteen hundred "qualities and states-of-being" pertaining to the legal system. Since the legal system impinges on other areas of life, Legal-RDF includes vocabularies for documents, contacts, economics, events, locations, measurements, products, and property. The section that attempts to provide a vocabulary for every possible legal event includes a subsection on actions that starts with "AbandonmentAct" and ends with "ZoningSite." Each is briefly defined and classified: "An 'AbandonmentAct' is a kind of AbandonmentEvent and DisposalAct." A competing legal ontology, LRI Core, is even broader, seeming to map all of creation, with physical, mental, and abstract concepts at its root level. Unfortunately, such projects re-create the same problems faced by the traditional categorizations of knowledge: Human topics are too big and squishy to fit well into any one set of boxes.

"It's not the Semantic Web's fault that some people are compulsive," says Tim Falconer, who makes most of his living mentoring companies about the Semantic Web. Falconer represents the other side of the dialectic, the other mood in the bipolarism of understanding. For him, the Semantic Web is about "smushiness." Fittingly, Falconer has trouble coming up with a precise definition of that term. He settles on "pragmatic looseness." "It's better to do something and tweak it for the rest of your life than to get thirty people into a room to figure out everything you're ever going to need," he says.

Falconer and other "smushies" agree with the Big Ontology folks that the Semantic Web is about using RDF triples to express the relationships you want to share with others. But Falconer doesn't think you have to specify an exhaustive set of relationships for them to start having value. Ontologies can be built bit by bit, reusing work that's common across domains. After all, the basic relationship that documents are written by authors holds whether you're a doctor, astronomer, baker, philatelist, or banker. So if you're designing a new ontology, rather than redefining the relationship of documents and authors, the smushy approach is to have your ontology refer to an existing document ontology, stitching together relationships already defined. It's messy, but it avoids the overweening task of plotting out all at once all the relationships implicit in a large domain.

While the smushy approach makes the Semantic Web more plausible by loosening the constraints, it doesn't satisfy everyone. For example, John Frank argues that the company he founded, MetaCarta, provides a perfect example of the inadequacy of the Semantic Web's approach. MetaCarta's software analyzes the language in documents, looking for references to geographic places. It tells the user that there is a certain probability that the word "London" in a document refers to London, England, another probability that it refers to London, Ontario, and a third probability that it refers to London broil. But RDF triples were not designed with probabilities in mind. They assume relationships are as simple as "is the daughter of." Yet even being the daughter of someone is more complex than those few words express. Just ask any daughter. While there are work-arounds to get RDF to express complex relationships, it was not primarily designed for a sort-of and kind-of world. RDF would have made Aristotle happy. It would not please Eleanor Rosch.

The witticism about why artificial intelligence has never lived up to its promise is that as soon as an AI application succeeds, it no longer looks like AI. The same is happening with the Semantic Web. Businesses are making use of it, but not in the grandiose ways sometimes promised. It holds promise in health care, where the ability to interchange information can improve the treatment of patients, and in health sciences, where pulling together data from the multiple Web-based databases of chemical information could prevent surprises about toxic drug interactions. NeuroCommons.org makes neuroscience information available online in Semantic Web format. The Air Force Research Laboratory is developing an ontology to help speed clearance for traversing foreign airspace. IBM is connecting the Life Science IDs (LSIDs) discussed earlier via RDF. But large-scale Semantic Web applications out of the research phase are hard to come by. At this point in the development of the World Wide Web, hundreds of millions of people were online and billions of pages had been created. In comparison, the Semantic Web is moving glacially—more on a *Britannica* schedule than at a Wikipedia pace. As the former chief information officer of Utah, Phil Windley, concluded at the end of a blog post about a panel Berners-Lee was on, called "The Next Wave of the Web," the Semantic Web is "*still* being talked about in the future tense and as something that 'will be really cool when it gets here.' "

Smaller-scale approaches are working. Microformats, for example, are a lighter and faster way to gain some of the benefits of the Semantic Web. Rather than waiting for industry giants to agree on the details of an intricate ontology, a small group of people can design a microformat that captures perhaps 80 percent of the metadata, so it's usable immediately, even if it's not perfect. For example, there's a microformat for the metadata associated with product and entertainment reviews that's been adopted by one of Japan's largest movie review sites, and a microformat for calendar events is used by Yahoo! and the 2006 Nobel Conference. But microformats conspicuously do not base themselves around RDF triplets. They're really just old-fashioned agreements about what metadata to capture.

A *world wide* Semantic Web is so ambitious that it falls prey to the same problems that beset Dewey and other large taxonomies. A Semantic Web that loosely stitches together imperfect, smushy, local efforts is not only more likely, it is to be preferred. A seamless whole that drives out ambiguity would also drive out the richness of implicit meanings. Ironically, then, the key to the success of Sir Tim's attempt to clean up the World Wide Mess may be that it needs to get good and messy.

THE SORT-OF, KIND-OF WORLD

In the early 1930s, S. R. Ranganathan could have had no idea that his Colon Classification system would have its most important effect only after computers were invented: the development of faceted classification systems that enable people to navigate through category trees constructed on the fly. Eleanor Rosch is likely to be in the same position as the Web takes what it needs from her theory. Prototypes are as hard to find on the Web as is Colon Classification, although search engines such as Google let users click on a page in a results list to find other items like it, and face-recognition systems are getting better at identifying pictures of Aunt Sally by first having us pick a prototypical image of her. While prototypes are unlikely to become a dominant way of organizing Web materials, the fundamental property of prototype theory is already quite important in the digital order: The Web is full of sort-of, kind-of clustering based on multiple attributes, not based on Aristotelian definitions.

"Something can be 73 percent in a category," says Joshua Schachter, the creator of Delicious. "The edges are fuzzy." Aristotelian trees and RDF triples have trouble accommodating that, but there is no other way to make sense of the "tagosphere," as some call the collection of publicly-available tags being generated at Delicious and Flickr. We've only forced ideas into unambiguous categories through authority and discipline. The folksonomies that are emerging bottom up are characterized by ambiguity, multiple classification, and sort-of kind-of relationships.

For example, when Flickr automatically splits photos tagged "Capri" into photos of the Italian island or of the Ford car, it also shows the additional tags associated—sort of, kind of—with each of the two groups. Thus, next to the Capri island cluster it lists "italy," "sea," "island," "water," "Italia," "blue," "naples," "Napoli," "Europe," and "boat." The first three in the list are in boldface to indicate that the statistical correlation is particularly strong—"73 percent in a category," in Schachter's terms. Likewise, if you browse all the photos at Flickr tagged "Italian," you'll see photos of Capri, the Colosseum, a plate of roasted pork loin on top of asparagus, an Italian plant manager in what seems to be a motorcycle factory, a red beverage, a high-voltage sign in Italian, and a glamour shot of a toothbrush loaded with toothpaste. Such a cluster of photos is not a true case of a family resemblance, because all of those photos do indeed have one characteristic in common: Someone has tagged them "italy." But, like a family resemblance, there is no single explanation of what makes "italy" an appropriate tag. It's obvious why the photo of the Tuscany landscape was tagged that way. We can guess why the photo of the pretty, dark-haired woman was tagged "italy," although we can't be sure if it's a photo of an Italian or of a visitor to Italy. As for the photo of the half-completed knitting project, we will probably never know how it's even sort-of, kind-of related.

These types of clusters have the properties of Roschian prototype categorization without there ever having been a prototype. No one, and no computer program, ever picked out a few photos, identified them as great examples of Italy, and then used them as the exemplars against which other photos are compared. But, as with a prototype categorization, the clustering is loose and a matter of degree. And, importantly, a photo can be in as many categories as anyone wants. In fact, instead of clusters forming around prototypes, prototypes could emerge from clusters: There are relatively few photos at Flickr tagged "italy" and "toothbrush," but many tagged "italy" and "rome," so it would not be hard for Flickr to isolate some photos as likely to be prototypical of "italy."

Clustering does not satisfy every need. Sort-of, kind-of relationships are not adequate for air-traffic controllers or brain surgeons. Nor is clustering the only way to bring order to the Web. As we've seen, because the Web is a third-order mess, every Oscar and Felix can order it however he wants. Smushies and system builders can build up a supply of triples without diminishing anyone else's way of ordering what's there. Sometimes triples will be just what we need. Perhaps we'll invent variants to make RDF better able to capture degrees of relationship. Every triple, every playlist, every hyperlink adds value to the mess. None diminishes that value because none actually cleans up the mess, just as uttering sentences does not use up language.

But if the third-order mess is like Heraclitus's river that's different every time we step into it, how can we *know* anything? Rosch's prototype theory suggests that if knowing the world means seeing the swarm of impressions divided cleanly by a single set of precise definitions, then knowledge is a diminished view of the world. Sometimes that's what we need: By defining the chemical elements rigorously by means of a single property (atomic number), we can build compounds that save our lives. Such stipulations can be crucial. But they are the exception, and they are not all that matter. In the sort-of, kind-of world in which a leaf can hang from many branches, our task becomes less to discover the one thing that something is than to see what it sort-of, kind-of, 73 percent is. The task of knowing is no longer to see the simple. It is to swim in the complex.

10

THE WORK OF KNOWLEDGE

Specimen No. 1212 is a 137-year-old gnat. David Turell, an imaging technician in the Entomology Department at Harvard's Museum of Comparative Zoology, is sticking a pin's extra-thin shaft through the gnat and a tiny red label, also 137 years old. The label is red because the gnat is a *type specimen,* the actual organism found by the naturalist who first identified the species. If two naturalists disagree about whether a particular gnat is *Glaphyroptera decora Loew,* the insect at the end of the pin in Turell's hand settles the issue. Without collections like this one and the one at the Linnean Society headquarters in London, the tree of species would have no roots.

The gnat can exercise its authority only because the number on the red label points to one of six ledgers kept in a drawer in Professor Philip Perkins's office, down the hall. Each line in the ledgers has one handwritten entry, beginning with *Cicindela amoena LeC.* (tiger moth) and ending with No. 35518, *Acanthoscelis griseus Dietz.* Perkins is the latest in the line of keepers of the ledgers and has recorded three or four hundred entries so far. Although he can tell by the handwriting where his contribution begins, he's not sure when that was because the entries do not contain dates—metadata that would be useful in a miscellanized world.

Specimen No. 1212 is a tiny, dried husk skewered by metal. Turell has withdrawn it from its protective box so he can take five or six high-resolution, close-up pictures of it. The cameras are digital, of course. "Some of these insects have never been seen fully in focus before," says Brian Farrell, professor of biology in the Department of Organismal and Evolutionary Biology, and Turell's boss. Because the cameras are able to focus on only one part of the insect at a time, software stitches together the in-focus portions of several images. Now we can see every pore, every whisker, every arthropodal fetlock in glorious detail, all at once.

Turell types "1212" into a computer on his desk and calls up the record for that specimen. It tells Turell that this particular gnat should be returned to Cabinet 26, Drawer 11. A second label on the insect shish kebab says that the gnat belongs to the Loew Collection, donated in 1869. Turell carefully sticks the pin back into the bottom of its small box. It will soon be returned to its metal tomb. The images will be posted on the Web, with multiply redundant copies. After all these years, the brittle anchor of our knowledge about this one species has a backup.

We're not at risk of turning away from knowledge in this new age of the digital miscellany. We are too good at knowing ever to let it go. Indeed, we're making knowledge our new currency. But everything touching knowledge and everything knowledge touches is being transformed. Traditional knowledge, like a lighthouse as the sea recedes and as radar supplements static maps, is changing simply by staying the same. Big questions loom:

As Umberto Eco says, there are many possible cuts of beef, but it's hard to imagine one that has the snout attached to the tail. Even so, if there are many ways to slice up the world, what happens if we don't slice it up the same way as others? *Is knowledge being fragmented? Are we being fragmented along with it?*

The miscellaneous is unowned. Anyone can add to it. Anyone can slice it up and reorganize it the way she likes. *What happens to the very notion of a topic when there are so many ways to carve up nature?*

Freed of paper, our knowledge can now be presented, communicated, and preserved in ways rich with links and exceptions. *Does knowledge stay simple and orderly?*

In the miscellanized world, knowledge is at most one click away from everything else that is not knowledge. Often they share the same page. *Does knowledge retain its privileged position?*

Finally, we can re-ask the topic we began with: *If everything is miscellaneous, why doesn't it stay that way?*

SHARD KNOWLEDGE

When Howard Dean's campaign for the 2004 Democratic presidential nomination suddenly failed, the metaphors used to describe it suggested that his support had been illusory. The bubble burst, the Democratic Party woke up. Observers wondered why anyone had ever believed that Dean was the leading candidate. For Clay Shirky—whom we've met as a skeptic about the power of top-down taxonomies—it was a "collective delusion" caused by supporters speaking only with those who agreed with them. The law professor Cass Sunstein had already worried, in his 2001 book *Republic.com,* that "an unlimited power to filter threatens to create excessive fragmentation." "For countless people, the Internet is producing a substantial decrease in unanticipated, unchosen interactions with others," he wrote. This fragmentation is not due only to the Internet:

> If you take the ten most highly rated television programs for whites, and then take the ten most highly rated programs for African-Americans, you will find little overlap between them. Indeed, seven of the ten most highly rated programs for African-Americans rank as the very *least* popular programs for whites.

The emerging evidence suggests, he says, that "many people are mostly hearing more and louder echoes of their own voices." Worse, according to Sunstein this fragmentation is causing groups—shards—to become more extreme and more polarized in their views.

The top-down map of the Web pictured by Albert-László Barabási in his book *Linked* seems to bear this out. It shows a relative few "hub" sites thick with connections, while most others have only a few lines going into them—like one of Valdis Krebs's drawings, except with an even greater centralization around a few key workers. This seems to prove that the Net has repeated the basic structure of the broadcast medium: a few speakers with lots of listeners. It seems the utopian vision of the Internet as a place where everyone gets to be heard equally is merely an empty promise made by woolly-headed thinkers and aging hippies.

But the Yale economist Yochai Benkler, in *The Wealth of Networks,* sees a different picture. Benkler says the right question isn't whether the Web provides perfect equality but whether it provides more equality than "the one-way structure of the commercial mass media." Through structural analysis and case studies, Benkler shows that the Web is far more complex than it seems in static maps. Typically, he says, loose sets of low-traffic, interest-based sites talk among themselves. If a topic develops that is of sufficient interest, it may be picked up by one of the larger "regional" sites that attract lots of traffic. This brings it to the attention of other local sites that share the regional sites' interests. It also may bring it to the attention of sites with contrary views, generating interest among the sites clustered around them. The path of ideas tells a story of constant conversation, elaboration, and disagreement that is not visible in a simple map of links, just as an organization's actual social networks are invisible on the static organizational chart.

Benkler reminds us that hyperlinks are not just paper clips joining two sites. If I link to your site, I probably do so with a comment such as "So-and-so makes some good points," or "Here's a product that's really different." That context, in which people link to the ideas and opinions they're writing about, adopting a "see for yourself" attitude, works against polarization, he says. (It works against the simplifications and overstatements of advertising, too.) As Benkler points out, each "node" on the map—think of Digg.com, composed of links nominated by readers, or a blog post with comments—may itself contain a cluster of contributors. It's all much messier than even the network maps make it seem, yet in that mess are conversations in which ideas reach "salience." That makes the Internet a powerful force for democratic institutions and open markets, not a polarizing and simplifying medium.

Web conversations have looked to many like echo chambers because of the nature of conversation itself. Conversation always occurs on a ground of agreement. If we don't first tacitly agree that sugar is an edible substance, we can't then talk about whether eating sugar makes kids crazy. From that basis of agreement, we then iterate on differences. Where the ground of agreement is more controversial—"Howard Dean should be president" at the Dean site or "Aromatherapy works!" at a New Age healing site—outsiders may think that a bunch of people who agree have gotten together to reinforce one another. But that mistakes the ground of conversation for the conversation itself. By discussing differences while standing on a shared ground, we work toward *understanding.*

Understanding, not knowledge, is what we're aiming at in most conversations. Philosophers have told us for a couple of millennia that knowing is the highest of human mental activities, but that's because you don't become a philosopher unless you're interested in getting past the mere opinions of those around you to find what's truly worthy of belief. It's like asking a chef which is the greatest of the senses or a libertine what's the greatest thing two people can do together. Philosophers after Socrates have tended to sit in a room by themselves as they write things on paper, but most of us think by talking with others. In a conversation's shared ground there are things we know—or assume we know—but they're precisely what's not interesting to talk about. In conversation we think out loud together, trying to understand.

The noise this makes is very different from the scratch of a philosopher's ink on paper. Paper drives thought into our heads. The Web releases thoughts before they're ready so we can work on them together. And in those conversations we hear multiple understandings of the world, for conversation thrives on difference. Traditionally, difference has been a sign that knowledge hasn't been reached: There can be only one knowledge because the world is one way and not any other. But there will always be multiple conversations and thus multiple understandings. We're never going to stop talking with one another, silenced by the single, unified, true, inescapable, and final knowledge of all that is.

Where was this unified knowledge before the world went miscellaneous on us? In the *Encyclopaedia Britannica*? The *Britannica* is a great encyclopedia, but it doesn't contain everything, it's not right about everything, the paper version is always out of date, and not everyone accepts it as an authority. Even within that great work, knowledge is not homogeneously authoritative. It's reasonable to take issue with the *Britannica*'s assessment of the role of Thomas Jefferson in founding the United States, but less reasonable to argue with its assertion that Jefferson died in 1826. Likewise, we're right to trust the facts on the specifications page of a product brochure, but we're also right to be skeptical about the manufacturer's claims that those specs make the product "the world-class best" of its type. Our knowledge becomes more unitary and undeniable the less interesting it is. Conversation, on the other hand, is always about what's interesting enough to justify the time and wind. From it comes understanding that can be true and useful even though we will never all understand our world in exactly the same way. What is a weakness in knowledge is a strength of understanding.

It can also be a political strength. The claim that the Howard Dean campaign was an "echo chamber" reframes negatively what was in fact a significant positive achievement. Joe Trippi, the campaign manager, explicitly set out to break the usual model, based on top-down control and centralized ownership of the message. That's how broadcast media work and it's been how political campaigns have worked, but when the Dean campaign started, it didn't have enough money to run that type of campaign. So Trippi assembled a remarkable staff of political organizers and Net innovators—including people like Zephyr Teachout, who combined both—and they set out to mix it up. Rather than treating supporters as foot soldiers marching to the beat of the daily message, the campaign encouraged and enabled supporters to find one another and affiliate based on shared interests, location, and plain old friendship. Rather than telling supporters what message to communicate, the campaign encouraged and enabled supporters to develop their own. For example, when the campaign took up a supporter's idea that fellow "Deaniacs" write to voters in Iowa before the primary there, it purposefully did not provide a template letter to be followed. The campaign's Web site featured a blog that sounded like it was written by a supporter given inside access because that's exactly what it was. There would routinely be four hundred comments on any single post, unfiltered and frank. All these ways of letting go by the campaign were rightfully taken as a sign of trust by the supporters and were rewarded with a loyalty and enthusiasm of which marketers dream. Dean lost because of policy, organizational, and personality reasons. That he got as far as he did is testimony to the power of the miscellaneous.

KNOWLEDGE UNCHAINED

Einstein was no Einstein when it came to world politics. At least, not necessarily. Genius is topical. It therefore has to be proved anew in every domain. We even have a fallacy with a Latin name—*Argumentum ad verecundiam*—to remind us not to think that just because a person is an expert in one field, she's also to be relied upon in other fields.

How big is a reasonably sized field for expertise? It's a silly question if we expect an answer with any precision, as if we could send out surveyors to mark the edges. But there are some informal metrics. For example, if a field of inquiry is too small to fill up a single book—*The Left Front Legs of Crickets*—it's probably too narrow to be an accredited field of study. If, on the other hand, the field has thousands of books within it—*Living Things, Volume 34,756*—we'll be skeptical of anyone who claims expertise that broad. The span of expertise is about as long as a shelf in a library.

It is no accident that books provide a way of talking about the size of fields of expertise. Books have to be *about* something just as an expert has to be an expert *in* something. Books, like experts, are valued because of the knowledge they contain—experts cover a topic the way a book fits the topic between its covers. Writing a learned book certifies the author as an expert. Students of Marshall McLuhan even argue—quite plausibly—that the modern notion of knowledge and expertise came about because of the invention of the printing press.

But what happens if topics crumble? What if knowledge doesn't divide into stable, mappable fields? What then will experts master? If masters no longer have a territory, what are they masters of?

The *Britannica* picks its topics carefully because it has limited space and because the encyclopedia's value comes from its careful editorial process. That forces hard choices. "Totemism is a subject of growing importance," wrote editor William Robertson Smith to an editorial associate while overseeing the ninth edition, completed in 1889. "We must make room for it whatever else goes." His suggested cut: torture. Wikipedia, on the other hand, does not have such constraints; it can always just add some more hard disks.

Wikipedia has an article that compares the sizes of various encyclopedias, including itself. In January 2006, it reported that the English-language edition had 1,407,237 articles, with 511 million words. That works out to 363 words per article. The *Britannica*'s 85,000 articles are on average 650 words long. Conclusion: *Britannica* articles are generally almost twice as long as Wikipedia articles.

But this comparison is too facile. The two works have very different approaches to divvying knowledge up into topics. In the 1970s, *Britannica* added a set of volumes, called the Macropedia, that deal with a few topics at great length. The combined word-tonnage of the Macropedia philosophy articles—from philosophical anthropology to the history of Western philosophy—is a whopping 184,800 words, the length of three midsized books. The philosophy entry in the "normal" part of the encyclopedia, called the Micropedia, boils philosophy down to a terse main entry of 279 words. The philosophy entry at Wikipedia, on the third hand, has 4,133 words, about fifteen times the size of the Micropedia entry but a mere one-forty-fifth of the Macropedia entry. Does that mean Wikipedia has less about philosophy than the *Britannica*? Not necessarily. Wikipedia's style guide suggests that articles not exceed 32KB, about 6,000 to 10,000 words, a principle originally instituted because some browsers couldn't handle files larger than that, though kept on now for stylistic reasons. Because of this, excess material is moved to its own article, with links from the original article pointing to it. Hyperlinks mean that Wikipedia doesn't have to bring everything it knows about philosophy under one topical roof. That relieves the pressure to get it all down in one spot. But it also fundamentally changes what constitutes a topic.

The loose linking of topics means that Wikipedia is a prime example of information sprawl, the natural topology of the miscellaneous. As we've seen, it has its advantages. Consider the response of the Cambridge don F. L. Lucas in 1961 when the *Britannica* instructed him to cut by half the entry on Oliver Goldsmith (written by the famous British historian Thomas Macaulay, no less):

> One's encyclopedias grow less useful, because what one wants to know is crowded out by things one doesn't want to know. . . . By A.D. 3000, no doubt, dear Oliver will be reduced to a couple of lines.

In the 1911 edition of the *Britannica,* the Goldsmith article was 6,000 words long. In the latest edition it's down to 1,500 words.

Wikipedia articles, on the other hand, tend to get longer. For example, the Micropedia has 248 words on Edith Piaf, but if you're enough of a fan of the French chanteuse that you're going to write a Wikipedia article on her, why would you stop at 248 words? And as other fans read it, they're likely to add more details about her life, her effect on music, her role as a cultural icon, perhaps a discography. At Wikipedia, topics assume their natural size.

Print not only forces editors to make unnatural decisions, it layers symbolism onto the length of topics. For instance, you can tell at a glance that the *Britannica* entry on Pekalongan, a municipality in Java, is a minor article because it gets a mere 109 words. If the Wikipedia Pekalongan article grows from its current 426 words into a 20,000-word essay with color photos and complete coverage of local politics, readers won't say—as they would if it were in *Britannica*—that the editors must either have lost their minds or invested in Pekalongan real restate. Instead, readers will think, "Gosh, there are people who must really love that place." In the *Britannica*, length is a symbol of importance. In Wikipedia, length is a manifestation of interest and passion, even if the interest and passion of only a single person. And while the length of any single topic at Wikipedia may not tell us much, Wikipedia overall does tell us that the world is more interesting than the *Britannica* lets on.

Wikipedia also shows us that topics are busting out of their bindings. The *Britannica* includes references at the end of articles to remind us that topics are related to other topics, literally afterthoughts. Wikipedia, on the other hand, is besotted with links:

> **Richard Henry Sellers** CBE, (September 8, 1925–July 24, 1980), better known as **Peter Sellers,** was an English comedian, actor, and performer, who came to prominence on the BBC radio series The Goon Show, before embarking on a successful film career.

These links are not even bread crumbs, for with two clicks we well may be going down a path no one has trod before and that no one anticipated. You can even click on a date or a year and find out that on September 8, 1331, Stefan Dusan declared himself king of Serbia and that Sid Caesar and Lyndon Larouche were born on that day in 1922. If we're not sure who these people are, their names are also hyperlinked. Why pick one tree when we can swing through the vines?

In the miscellaneous order, a topic is anything someone somewhere is interested in. Anyone can pull a topic together by contributing to Wikipedia, writing a blog post, creating a playlist, or starting a discussion thread. Even loosely defined topics will typically be shot through with links leading us away, miscellanizing them. Topics lose the borders that make it easier to know when we've mastered them, and they also lose some of the dignity we've imposed on them. Jimmy Wales points to his own favorite Wikipedia category: fictional pigs, a subcategory of fictional animals. "There are a surprising number," he says, rattling off a few, from Snowball in *Animal Farm* to Wilbur in *Charlotte's Web*. "I think that stuff is a hoot."

That's one final characteristic of the miscellanizing of topics. No one ever said that the *Encyclopaedia Britannica*'s topics themselves are a hoot.

COMPLEXIFIED KNOWLEDGE

On May 15, 2006, President George Bush addressed the nation. In 2,537 words, he laid out the problems raised by illegal immigration, proposed a solution, and exhorted us to accept it.

Just a few hours later, over 2,400 bloggers had commented on that speech—about one post per word in the president's address. Mike Beattie at Purely Random pointed out that while Bush was governor of Texas, he had a history of supporting immigrants. The Blast Furnace Canada Blog reported that California Governor Arnold Schwarzenegger had been critical of Bush's speech on immigration a year earlier. The PEEK blog noted that it was the first speech on domestic policy that Bush had given from the Oval Office. Blogs typically pull out an implication, draw an unexpected conclusion, or connect an idea with another idea. Thus did the blogosphere undo the careful work of the president's advisers to make the big hairy problem of immigration simple and clear. Simple arguments, simple ideas, simple language. That's how politicians talk. But it's not how we, their constituents, talk.

Marketers also want to simplify our world for us. At its Web site, the Siegel+Gale marketing firm announces, "simple is smart," without any capitals or punctuation to complicate matters. Its site brags that they helped Lehman Brothers boil itself down to "Where vision gets built," Lexus to "Making the Most of Every Moment," and Berklee College of Music to "Nothing Conservatory About it." "Simplicity, simplicity, simplicity!" wrote Henry David Thoreau. But he embedded that slogan in a complex book, and refrained from repeating it in radio spots aired eighteen times a day across the nation.

Marketers now have to compete with the conversations customers are having with one another about the products they buy. None of those conversations consists of customers repeating the same three-word phrases over and over. This is one of the main drivers for marketing's interest in "customer-generated media": Not only are customers more credible—a 2006 study by Edelman PR showed that customers think the most trustworthy source of information about a company is "a person like me"—they're also more interesting. Customers now are "mashing up" marketing materials—re-editing them into parodies, mixing them up with totally inappropriate soundtracks—turning commercials back against their creators and in the process making them far more interesting than they were originally. Think of it as customers' revenge for all those years of being treated like simpletons.

Science, despite its complexity, is also in search of the simple. In the Istituto e Museo di Storia della Scienza (the Institute and Museum of the History of Science), just a few blocks away from the river Arno in Florence, sits an armillary sphere built at the request of Ferdinand I de' Medici between 1588 and 1593. At its center is the Earth, around which are wrapped rings and spherical bands connected by gears and corkscrews in a surprisingly asymmetrical fashion. Turn the machine's handle with sufficient force, and this clockwork universe moves the heavenly bodies accurately across the sky as seen from the Earth, tracing the seemingly irregular movement of the planets by describing circles rotating in circles, like twirling teacups in an amusement park ride. The complexity of the armillary's mechanism springs from a commitment to using only circles—considered by the ancients to be the simplest and most perfect of movements—to describe the cosmos. When Johannes Kepler finally realized that the planets and the earth move around the sun in elliptical orbits, a single formula suddenly sufficed to explain what had required a room full of gears. The heavens got simpler. Science sighed in relief.

Science, of course, is not simpleminded. Nevertheless, in finding single causes for multiple events (a particular law explains every planet's motion, a particular germ is behind multiple instances of a disease), science finds an explanation that's simpler than what it

object before us with confidence that the complexity is there if we need it. But trees favor simplicity: A leaf can hang from only one branch. All the relationships among the branches are the same. One particular attribute is given priority: Bananas in the tree of food inherit "fruit" from the branches above it, but not "yellow" or "phallus-shaped." The leaves are treated as fundamentally discrete when in fact they may be multifaceted or impossible to define precisely. Even the basic notion of containment expressed by a tree's branching structure is way too general. Does "color" contain "red" the way "nation" contains "city" and the way "actor" contains "David Caruso"? Does "pants" or "shorts" contain "Capris"? And "yard" does not contain "dog" even if your dog is in your yard, and "stomach" does not contain "peanut" even if you've just eaten one. Despite Aristotle's principles, knowledge is not shaped like a tree. It seemed that way when we were lumping and splitting ideas the way we lump and split laundry and the way we stock the shelves of our stores. But in the third order, we can tag ideas in as many categories as we wish. Something can be 78 percent in one pile, 63 percent in another, and 54 percent in a third—or a potential team member can be a *pretty good* French speaker, a *great* applications expert, a *mediocre* people person, and *very reasonably* priced. In the sort-of, kind-of world, the percentages don't even have to add up to one hundred.

The difference in the digital order is the difference between the annoying interactions you have on a product support line—"Press 1 if you're calling about a medical emergency. Press 2 if you're calling about billing"—and the conversations you have with real people. Maybe you and I start out talking about our sons' asthma, and before you know it, we're laughing about our first pets. Sometimes those connections will be nothing more than entertaining, but sometimes we'll discover that a cat allergy may explain why we can't sleep through the night. The potential for connections from the trivial to the urgent is characteristic of the new miscellany. We are busily creating as many of these meaningful connections as we can.

Because we are doing this willy-nilly and sometimes without even intending to, we are blurring lines faster than we draw them. This phenomenon is familiar to us. At first a box of eight crayons was enough to capture our world, but over time we needed to mix our own colors. We learned to see past the reductions and stereotypes in every field. In the miscellanized world, every idea is discussed, so no idea remains simple for long.

THE PLACE OF KNOWLEDGE

If we are defined as the animals that are rational, then knowing is the highest human activity and knowledge is king. But the third order of order doesn't have a lot of patience with monarchs who tell us how we shall organize our ideas. The fate of the king rests on three questions: What's happening to the knowledge we already know? What's happening to how we develop knowledge? And what will be knowledge's role in the externalized web of meaning we're spinning?

The Knowledge We Know

In some conceivable world, the denizens created an Internet filled with nothing but factual information. They can find the annual revenues generated by a particular model electric keyboard but not download an amazing recording made by a twelve-year-old prodigy using one. They can look up the bus schedules but not hear darling anecdotes about what some child said to some old man on the No. 66 the other day. They can read their political parties' platforms and the precinct-by-precinct election results but not what a Pennsylvanian Sufi lesbian tinsmith with a Ph.D. in agronomy thinks about the candidates' boutonnieres. In this imaginary world, the Internet is a web of facts and nothing but the facts, and Sergeant Joe Friday rules.

There are, of course, facts on our Internet, too. Many Wikipedia articles about famous people include a box near the top that contains the basic biographical facts, just as newspapers include boxed summaries of information about countries they consider obscure to their readers. And while you would be wise not to believe a typical business site when it tells you that its products are "whisper quiet" or "tough on dirt," you should generally believe facts in boxes—including the dimensions and materials the company puts in a box labeled "Specifications." We set them apart because they are the indisputable part. Facts are that about which we no longer argue.

We don't *always* agree about facts. But the prototype examples—in Eleanor Rosch's sense—of facts are simple statements about measurable quantities about which all reasonable people agree: the speed of light, but not the world's sexiest man; the tallest mountain, but not the longest short story; the cat is on the mat, but not cats are better pets than dogs. If there's serious debate about an issue, then we say we don't yet know the facts. Facts, once established, are like commodities—the products that are so widely available and of so little distinguishable value taken one by one that suppliers can sell them only at very low prices. To a hardware store, nails are commodities, but power tools aren't.

Commodities are important. Our civilization would crumble if all the nails were removed. Even more would fall if all the facts were extracted. Yet for $9.97 you can buy the latest *World Almanac and Book of Facts,* 1,012 pages of small-print facts. Bill McGeveran, who had retired as editorial director a few weeks before I spoke with him in 2006, guesses that "there must be hundreds of facts on a page." Facts don't differentiate almanacs, any more than nails differentiate hardware stores. "All the almanacs are going to get the current gross products of the states," McGeveran says, "and it'll be the same in all of them, assuming all close at the same time of year. But if you break median income down by race, do you want to see every year? Or five- and ten-year trends?" That's the type of editorial decision that determines which almanac you're going to put on your desk. Likewise, Wikipedia puts into the George W. Bush fact box that he was born on July 6, 1946, and that he's married to Laura Welch Bush, but not who his parents were, what religion he practices, or what country his ancestors came from. One culture's fact box is another's trivia.

Now imagine it is ten years from now. New topics are still being added to Wikipedia and old ones edited, but not at the rate of the early years. The big arguments have mainly been settled. There are continuous small edits polishing the more popular articles, but big changes have become more rare. Wikipedia then constitutes the body of knowledge about which we agree. Wikipedia is doing to knowledge what almanacs do to facts. Wikipedia is commoditizing knowledge, continuing a trend that search engines such as Google began. Textbooks also present settled knowledge, or at least present it *as* settled, but the Internet makes knowledge as instantly available as a calculator's "equals" button.

Not all knowledge will be commoditized. There's always going to be plenty to discover and to argue about. And there will be localized knowledge commodities, the equivalent of "The Liberal Democrats' Wikipedia" and "The Lord's Wikipedia." But the commoditization of knowledge shifts the value proposition elsewhere in the value chain, as business folks like to say. *The World Almanac* was founded in 1868 so reporters, with the facts at their fingertips, would be freed to work one level up the value chain, writing articles that rely on those facts. In the same way, the commoditization of knowledge frees us to understand. Generally we understand something when we see how the pieces fit together. Understanding is metaknowledge.

The commoditization of knowledge enables greater value to be built from it, just as commoditized nails and lumber let us build better family homes for more people. But now more than ever, knowledge's value will come from the understanding it enables.

And since the commoditization of knowledge includes its easy accessibility, business loses one of its traditional assets. Information may not want to be free, in Stewart Brand's memorable phrase, but it sure wants to be dirt cheap. The good news for customers is that the miscellanized, commoditized knowledge sparks competition and innovation. The good news for businesses is that they can focus on providing the goods and services that are at the heart of their value.

Developing Knowledge

X-rays, the shape of DNA, the ozone hole, and the birth of Dolly the cloned sheep were all announced first in *Nature* magazine. *Nature* got to be the place scientists go to with major discoveries by patiently, issue after issue, putting articles through a process of anonymous peer review, allowing recognized experts to weigh in on the worth of each submission.

"I wouldn't imply that what we're doing is perfect," says Philip Campbell, the journal's editor in chief. There are problems with any peer-review process. Sometimes frauds slip through, and the system can create and perpetuate an orthodoxy. So in June 2006, *Nature* began a three-month experiment in which authors could agree to have their submissions posted for open comment, although the comments had no effect on which papers were accepted for publication. Campbell lists some other ideas the editors have been discussing with various degrees of seriousness, including providing each published author with a blog where readers can comment. "You could imagine a process at the end of which you're turning a paper into an open-source piece of work," he says. But, he is quick to add, "We will certainly continue with what we currently do." Why? Because anonymous peer review works. It distinguishes bad science from good, trivial reports from the important, shaky evidence from the reliable. It gives authority to what *Nature* publishes.

And, through a startling and persistent coincidence, all the knowledge developed in the natural sciences since 1869 has fit exactly into the number of pages allotted for it in *Nature* each week. Not one page more or one page less.

Of course not. *Nature* doesn't say how many submissions it gets every year, but its rough equivalent in the United States, *Science,* accepted less than 8 percent of the 12,000 articles it received in 2005. We know more than that are worthy of publication because many of the rejected articles are published in other prestigious journals. The knowledge published in *Nature* is determined not only by its rigorous peer-review process but by the economics of paper. Paper limits knowledge to what happens to fit into an object folded around vertical staples.

At a site called arXiv, we can see what knowledge looks like when the paper handcuffs are removed. Since 1991, physicists, biologists, computer scientists, and mathematicians have posted not-yet-published papers there—about 40,000 new papers every year, read by 35,000 people every day. (*Nature* has a circulation of 67,500 and claims 660,000 readers—about nineteen days' worth of arXiv's readers.) Many of the papers are later published in peer-reviewed journals, but all are available as soon as they are posted to anyone who wants to read them.

But we should be careful not to think that this is a battle royal between *Nature* and arXiv, paper and digits, top-down and bottom-up, filtered and raw, hidebound editors and freewheeling hippies. The third order is an ecology with niches of every sort. What starts out in the third order as open, authority-free, and permissionless can find itself evolving in unexpected ways. *Nature* encourages its authors to post their papers onto arXiv six months after they're published. And in January 2004, arXiv started requiring papers to be "endorsed" to be accepted into the archive. An endorsement relies on other arXiv authors or an academic affiliation to verify that the person submitting the paper has standing as a scientist and is not a hoaxer or a nutcase. The evolution of a site into a new niche can even occur off the site. For example, arXiv allows authors to include a brief comment on their own paper—Christopher Fuchs semifamously described one of his papers as "59 pages, 5 figures, 140 equations, one simple idea"—but it provides no way for readers to comment. So an independent site, Reddit.com, added its own ranking and commenting system for arXiv papers, without arXiv's knowledge or permission.

The Public Library of Science occupies other niches in this ecology. It was started by editors at peer-reviewed journals, including *Nature,* who want to put more research into the public domain. Hemai Parthasarathy, the managing editor of PLoS Biology, says that "Instead of trying to determine the top .001 percent of papers, it aims at publishing maybe the top 1 percent." PLoS Biology doesn't run all the scientifically sound research that survives its peer review process because, by being selective, the journal builds a reputation for quality that, in turn, attracts more high-quality research papers. But, acknowledges Parthasarathy, PLoS has an "intrinsic tension" because most of the editors there don't believe in "elite publishing." So they have started PLoS One, which publishes any paper that a peer review process determines is good science, no matter how important it's deemed. How will readers find articles? Will there be any guidance about which articles are especially worthwhile? Will they allow researchers, reviewers, or commenters to use pseudonyms? All these questions are to be decided, or, more precisely, the answers will emerge as the ecology of scientific knowledge selects the fittest solutions.

As ArXiv and *Nature* evolve, they smudge a line we once thought was clear. Just as we've thought that a statement is either true or false—per Aristotle's law of the excluded middle—we've thought something either is knowledge or it's not. But in the miscellaneous world, knowledge comes in gradations and varieties. Some knowledge is good enough to pass the most rigorous of peer reviews and make it into the pages of a prestigious journal. Some that pass peer review turn out to be well done but wrong. Some knowledge is reliable and important, but just not interesting enough for the top journals, so it shows up elsewhere. Some knowledge is unpublished but worth reading and discussing. Some knowledge is tantalizingly possible. Some knowledge used to be true, and some isn't true yet. If knowledge is king, the royal bloodline isn't as pure as we once thought.

Niches in this new ecology are distinguished by their metadata: We know an article in *Nature* is reliable because that's a peer-reviewed niche. Without metadata, we would be faced with an endless, indistinguishable ocean of articles. Because we're so good at handling metadata, we can get value from an unreviewed article at arXiv while knowing that it lacks confirmation. The metadata is a crucial part of the knowledge: This belief is rock solid because it's based on authoritative peer reviews, that belief is worth investigating because the evidence at nonreviewed sites is convincing, that other belief is required simply if we are not to despair.

Knowledge was supposed to be a mirror of reality. It thus was either true or not true, end of story. But if knowledge includes metadata about how much and why we should believe it, it's more like a mirror into which a teenager gazes, trying to figure out how she looks to other people. Knowledge can't be a literal read-off of the real because we're too deeply involved in the world we're trying to know. And just as we seem to be wired to recognize faces and emotions, we seem to be set up—by language and culture if not by neural anatomy—not only to know our world but to gauge the certainty and seriousness of what we know, whether it's a peer-reviewed science article, a product warranty with legal standing, or a thirty-second spot that we can't possibly take as seriously as it wants us to. As a species, we're born ready to grasp metadata. Our knowledge of the world is an understanding that simultaneously assesses the quality and reliability of our understanding.

Knowledge, Essence, and Meaning

In both British English and American English, the two most common words are *the* and *of*.

The identifies something as uniquely what it is. *Of* relates it to something else. It may seem at first that *the* is a word of separation and *of* is a word of connection, that *the* divides the world into neat units and *of* messes it up. But there's a hidden *of* in every *the*. *The* robin in your yard is only recognizable because it's a type *of* bird and perhaps a harbinger *of* spring. That was Aristotle's startling discovery: A thing, standing on its own, is what it is because of its connection to other things like it and other things not like it.

When we thought, as we did for a couple of millennia, that those connections were simple, elegant, and knowable by a rational inquirer, the first job of knowledge was to discern the defining criteria. Essentialism, taken at its simplest, says that each thing has a set of attributes that defines it, as well as less important attributes that come along for the ride. Rationality is part of the human essence, whereas the fact that our noses face forward is not. In classic essentialism, the essential definitions are perfectly knowable and arrange themselves in a neat tree with no overlaps, no gaps, and no exceptions.

We need look no further than *The Oprah Winfrey Show* to see how weak essentialism has become. In 1997, Tiger Woods responded to

Oprah that, yes, it did bother him to be thought of as simply African-American. "Growing up, I came up with this name: I'm a 'Cablinasian,'" he said, running together Caucasian, black, Indian, and Asian. Woods's genealogy is actually more mixed than that: His father is half black, a quarter American Indian, and a quarter white, while his mother is half Thai and half Chinese, making him more a "Blamincauthaichin."

Race was so important to us that it was one of only three questions on the very first census, in 1790, designed with the help of Thomas Jefferson. Yet for an "essential" category, race is surprisingly mutable. In the 1890 census, Tiger Woods would have had to choose between "mulatto," "quadroon," and "octoroon," depending on the percentage of his white ancestry. The 1990 census would have forced Woods to be a member of the race called "other," because he didn't fit into any of the other five pigeonholes: black, white, Asian and Pacific Islander, or Native American. Asian Indians were counted as white in the 1970 census but were counted in the Asian and Pacific Islander category in 1980. Hispanics were added in 1970 by President Nixon, under political pressure so intense that the forms had to be recalled from the printer to make the change. In October 1997, the Office of Management and Budget issued Statistical Directive 15, allowing respondents to check more than one box—a leaf on many branches—and write in "Hispanic" as their ethnicity, boosting the United States from five races to 126. Such distinctions are important; billions of dollars of government budgetary disbursements (let alone marketing researcher salaries) depend on them. And we citizens are no less confused than our census forms: Only 40 percent of those who declared themselves multiracial in the 2000 census said they were multiracial when asked in follow-up surveys.

Race, once considered so natural and so important that it determined if you would be a slave or free, "has no scientific justification in human biology," the American Anthropological Association said in 1997. That conclusion was based in part on an influential paper in the journal *Science* in 1972 that showed that members of a race differ among themselves genetically about as much as they differ from members of other races. You might as well say that men with male-pattern baldness constitute a race. Likewise, we used to have two sexes. Now, although gender still counts, especially when you're shopping for Speedos, knowing the shape of one's genitalia tells us less than we ever thought it did. As with skin color, that particular attribute no longer seems as defining as before.

Essentialism is failing in every way it can:

- Differences blend and gradate (Blamincauthaichin).
- The attributes we choose for dividing up the world depend on our assumptions and interests (skin color but not baldness).
- Attributes with real differences often turn out not to determine rigidly which other attributes come along for the ride (genitalia).

Some well-known categories fail all these tests: Planets are chosen by arbitrary characteristics that are less clear-cut than we'd thought and that don't matter anyway. This isn't to say there are no differences between men and women, that we need never consider race, that puppies are the same as potatoes. Rather, which differences we attend to has everything to do with our history, our language, and our intentions . . . and even then, the divisions are unlikely to be as clean as essentialism assumed they should be.

The postmodernists have been telling us this for a generation, as have the cognitive psychologists influenced by Eleanor Rosch. Antiessentialism was even a hot topic for John Locke in the late seventeenth century. That humans play a role in categorizing the world is not news. There is a difference now, though. For the first time, we have an infrastructure that allows us to hop over and around established categorizations with ease. We can make connections and relationships at a pace never before imagined. We are doing so together. We are doing so in public. Every hyperlink and every playlist enriches our shared miscellany, creating potential connections that we can't often anticipate. Each connection tells us something about the connected things, about the person who made the connection, about the culture in which a person could make such a connection, about the sorts of people who find that connection worth noticing. This is how meaning grows. Whether we're doing it on purpose or simply by leaving tracks behind us, the public construction of meaning is the most important project of the next hundred years.

We're going at this project at a devilish pace already, and it's only going to accelerate. Imagine, for instance, when electronic books are cheap and high-quality enough to begin displacing printed books. Every time a student highlights or annotates a page, that information will be used—with permission—to enhance the public metadata about the book. Even how long it takes people to get through pages or how often they go back to particular pages will enrich our third-order world. We'll be able to ask our books to highlight the passages most often reread by poets, A students, professors of literature, or Buddhist priests. Add in hardware that knows where books are being read and we can compile a playlist of beach books or travel books. We will be able to see what books our town is reading and which books our town has abandoned halfway through. Reading will cease being a one-way activity. It will become as social as the knowledge our children are developing as they instant-message one another about homework. All that metadata, and every use of metadata, will enrich the context within which we make sense of what we read and learn.

In the world after the Enlightenment, the cultural task was to build knowledge. In the miscellaneous world, the task is to build meaning, even though we can't yet know what we'll do with this new domain. Certainly some will mine it for knowledge that will change our lives through science and business. But knowledge will be only one product. Knowledge's new place will be in an ever-present mesh of social meaning. Knowledge is thus not being dethroned. We are way too good at knowing, and our continued progress—and survival—depends on it. But knowledge is now not our only project or our single highest calling. Making sense of what we know is the broader task, a task for understanding within the infrastructure of meaning.

META-BUSINESS

When businesses first began to take the Web seriously, the talk was all about "disintermediation." The Web would get rid of the people in the middle so customers could reach directly into product warehouses to get what they want. And, wherever a business process could be replaced by the equivalent of an ATM, it has happened: Customers now book plane tickets without consulting travel agents and buy music without physically thumbing through CD cases. But it's only happened where what appeared to the business to be its "added value" turned out to to be mere inefficiency in the customer's eyes. The products that got converted into commodities are the ones for which a business adds so little value that customers buy based primarily on price. But, in the miscellaneous world, information and even knowledge itself becomes commoditized. And that is changing the most basic, defining characteristic of business: Who owns what.

When you step into a physical store, you are entering territory owned and controlled by the store. The store manager lays out the merchandise in ways that help the customer get in and out quickly or that force the customer past enticing distractions. Either way, it's *the business's* choice because it's *the business's* store. That's how we divvy up the world of atoms. Yet, most business Web sites, no matter how many Flash animations and interactive buttons they have, continue to operate based on the most fundamental second-order principles. When you go to a commercial Web site, the business owns and controls the *information* it wants to give you, the way you'll *navigate* through that information, and the *experience* you'll have while doing so. If there's an email suggestion box on the site, the company thinks it's being open-minded.

The miscellanizing of information, knowledge, and ideas rips these assets out of the hands of individual businesses. Miscellanized information is information without borders. That means we've been misleading CEOs for the past fifteen years by drumming into their heads that every business is an information business. Of course information is central to businesses, but business's reflex action has been to wall off what they know as if it were gold. Now that information is being commoditized, it has more value if it's set free into the miscellaneous. For example, airlines do better when their proprietary scheduling and pricing information is made available to travel sites such as Expedia, Travelocity, and Orbitz. It gains even more value when innovators can combine it with other data, plotting it on maps, mashing it up with streams of ecological research, and plotting it against global economic trends.

Companies will try to hold onto some of their information, as is only proper. But there is an inevitable compulsion in the two imperatives of the miscellaneous: Thou shalt include and postpone. Including sometimes means pulling marginal information into one's own collection of data. But, with so much information to connect to, often it's more practical to leave the information where it is and to link to it. The miscellaneous is a distributed pile, pegged together through unique Web addresses, unique IDs, and good guesses about what relates to what. Just as the Internet itself could only scale by making it dead simple for one network to connect to another—it is an *inter*-network, after all—so too the miscellaneous can only scale by enabling local collections to make sense to other collections. As information gets commoditized and pulled out of a business to join in the general miscellany—where it is put to work and thereby benefits the company—the company is emptied of exactly what it's been told is its key asset in the new millennium.

At the same time, the miscellanizing of information is giving rise to a new category of business that enhances the value of information developed elsewhere and thus benefits the original creators of that information. Think of it as *meta-business.*

The rise of meta-business reverses the early expectation that the Web would disintermediate business by providing customers with direct access to goods. The Web has indeed cut out middlemen who provide no value, but it also provides an opportunity for new information-based businesses to emerge. The recording industry is the most obvious example. Record labels benefit as the information they develop about the products they sell is sucked into iTunes.

iTunes makes that information more searchable, more findable, and more usable by customers. Every playlist created by a user and published by iTunes markets record-label tracks at no cost to the label. But there's always the possibility of adding more value to information, so along come sites such as Pandora.com and Last.fm that create individualized "radio stations" for users, based on what others with similar tastes have liked and what songs are mathematically similar. That extra metadata introduces users to songs they might never have stumbled across, with a far greater likelihood that users will like what they hear—and perhaps purchase the track, go to the concert, wear the T-shirt—than if they were to tune into a real-world radio station trying to satisfy the tastes of an entire city.

Meta-businesses are arising across the board. Sites such as Expedia, Travelocity, and Orbitz don't merely list flight schedules, they let us see how flights compare by cost, departure time, departure city, airline, travel time, and more. Then they bundle flights into travel packages, driving business to rental car companies, hotels, and whitewater-rafting guides as well as back into the airlines themselves. Now other sites have arisen, taking the travel sites one meta step further. Kayak.com and FareCast.com include flights from low-cost travel packagers and provide users with more ways of sorting information—fulfilling the mandate to include and postpone. They also go up a level in metadata, displaying charts of how fares for particular itineraries have varied, and add information such as the average travel delays at specific airport terminals. The availability of this type of information may force the airlines to compete more fiercely, but that's how markets stay efficient and healthy.

Retail is also going meta, and not only with the sites that, from just about the beginning of the Web, have gathered commoditized information to let users compare prices, specifications, and customer service. DPReview.com has *more* information about cameras than the camera companies themselves do. Does the battery last as long as the camera manfacturer says? Does the automatic white balance work at high shutter speeds? DPReview provides context that the manufacturers prefer to skip, such as exactly what the problems were that caused Canon to introduce a new dust-clearing mechanism. New sites such as Wize.com take it to another level of meta-ness by aggregating links to customer and expert reviews written on sites such as DPReview. Wize uses this information to compute a meta-score for the product, but—in proper miscellaneous fashion—it lets users see which reviews went into that number and enables users to read the original reviews just by clicking.

The news industry is going to have to shape itself around the rise of meta-businesses, although it's far from clear what that industry will look like as a result. News aggregators such as Google News do the basic meta work, listing thousands of separate articles about the day's major stories. Digg.com and Reddit.com take the next meta step by using the collective wisdom of their readers to determine which stories are major. In November 2006, Reddit was acquired by the online arm of Condé Nast, a major content producer. Assuming that Reddit's new proprietors enable it to proceed along the path it was on, the acquisition implies that Condé Nast is undoing its most basic assumptions about its business: enabling users, not editors, to decide what's worth reading and providing a site where Condé Nast's own content has no special privilege. Only time will tell whether Condé Nast succeeds at Reddit's mission, but if it does not, others will.

The meta move presses upon every business that has information for users. Real estate customers are migrating from online MLS sites to sites such as Zillow and PropSmart that not only aggregate listing information but mash it up with other data. The automobile industry went meta long ago as sites quickly emerged that add value to each company's data simply by putting them up next to other companies'. Of course, the pressure to go meta increases when the industry's product is itself digital: Google's 2006 purchase of YouTube for $1.6 billion indicates how much value there is in aggregating content and providing new ways for customers to sort and organize it. And, as one meta business succeeds, others emerge to take it up a level. Dabble.com, for example, aggregates videos across YouTube and its competitors, enabling users to rank them and create playlists.

Meta business is inevitable because it adds value to information, and for that there will always and ever be a demand. And those businesses that are unwilling or unable to go meta—the CIA, for example, can't allow its information to be pulled into the public pile of miscellaneousness—may find themselves competing with companies whose information has been made more valuable, useful, and meaningful.

Going meta does understandably scare many traditional industries. The miscellanizing of information means that information is plucked from the tree of its birth and is available to anyone who can make use of it. It becomes more authoritative precisely because it's *not* on the site of the business that produced it. That means the originating business site doesn't have an opportunity to show off its carefully engineered customer experience, because the customers are going elsewhere.

This is, of course, also an opportunity. Google, the most successful business in the history of the Web, owns the information it has gathered (or at least the metadata it has gathered from the pages it has indexed), the ways to navigate it, and the experience users have on the site. But Google has been innovative in letting that information be miscellanized. For example, by making it easy for people to do mashups with Google Maps, combining maps with other information, Google maps have become the de facto standard on the Web. This easy integration of applications typical of what's been called "Web 2.0" allows information and services to be placeless, rather than locking them into the creator's site. Flickr has done the same with its collection of digital photos; all it asks from the user is a credit line (what we might call "advertising").

It may seem crazy to let other applications use the information and capabilities you've invested in developing, but it's often a generosity that pays itself back not only by introducing your product to new users but by making your product an integral part of their daily lives. It also says something important, something virtually impossible to say in the second order: Our business is truly all about you. Any marketer with a drop of sense would have advised Google when it was starting out to fill up the white space on its home page with ads, offers, and messages. But the fact that Google did not tells users something: There's nothing at Google that isn't about what helps the user. Likewise, when Wikipedia posts a notice that an article may not be neutral or accurate, it's telling its users that Wikipedia is dedicated above all to educating the user. This type of body language is so very different from the predatory crouch we encounter at most Web sites, where the obsessive aim is to get us to spend more than we intended. In the age of the miscellaneous, when we don't have to enter the lairs of predators to fetch the information we need, when that information is enhanced by being mixed up with other information and by being made more searchable and browsable, when we'd rather have the information, navigation, and experience anywhere *but* the company's Web site, the most successful businesses will have to get over the second-order assumption that they own the customer's experience. In a truly miscellaneous world, a successful business owns nothing but what it wants to sell us. The rest is ours.

WHY ISN'T EVERYTHING MISCELLANEOUS?

If everything is miscellaneous, why didn't it stay that way?

At the beginning of this book, I offered an answer: We work damn hard at straightening it up. We have built ornate systems of categorization that try to put all items in a domain—books, species of animals, photos, legal terms, employees—in their place. We have developed principles of categorization that not at all coincidentally mirror the limitations on lumping and splitting physical objects. We have built institutions that depend on maintaining systems of categorization for their authority and revenues.

With the rise of the third order of order, we can ask the question again. Why isn't everything miscellaneous? For, we do not spend our days swimming in Heraclitus's river, unable to tell if a robin is a bird or a handsaw.

The world and our third-order understanding of the world are miscellaneous in different ways. The world offers an indefinite number of joints without any preference about which ones we attend to: The rocks will continue to circle the sun whether or not the International Astronomical Union decides to stop calling some of them planets. The miscellaneous digital world we're building for ourselves, on the other hand, consists of what we have chosen as leaves—*Hamlet,* a particular edition of *Hamlet,* or a quotation from *Hamlet*—and the connections we've made explicitly or implicitly.

We inevitably make sense of what we experience. But the shape of sense is changing. We used to think ideas were well ordered when each was in the box that expressed its essence and the boxes were arranged neatly and elegantly. The world abetted us in this. Attributes tend to come in predictable bundles: Melons that smell good when their ends are squeezed tend to taste good, and animals with feathers and two feet tend to also have beaks and wings. We can cluster items by some of their attributes and reliably have other attributes come along. These separable but related traits and attributes are the real joints of nature, to use Plato's phrase one last time.

Over time, we've learned to undo some of these bundles of attributes, usually in thought but occasionally in matter. Sometimes these changes were forced on us, as when in the nineteenth century scientists were reluctantly brought to acknowledge that the platypus could have all the attributes of a mammal and yet lay eggs. But the requirement that we write things down hampered our ability to deal fluidly with attributes and categories. Paper's physicality dictated that topics had to be separable and confined to what fit between covers. Paper's immutability implied a fixity of knowledge. Paper's solitude lent itself to individual authors writing in locked rooms.

Freed of paper, we will continue our long march of knowledge, for we do it with uncanny skill. But in the third order, we turn an item over in our hands, noticing its glint and texture, trying to remember what it reminds us of. We make a note. The note is a public link that exists in the world and can be discovered and reused. The result is a startling change in our culture's belief that truth means accuracy, effectiveness requires adherence to clear lines of command and control, and knowledge is power.

It's not who is right and who is wrong. It's how different points of view are negotiated, given context, and embodied with passion and interest. Individuals thinking out loud now have weight, and authority and expertise are losing some of their gravity.

It's not whom you report to and who reports to you or how you filter someone else's experience. It's how messily you are connected and how thick with meaning are the links.

It's not what you know, and it's not even who you know. It's how much knowledge you give away. Hoarding knowledge diminishes your power because it diminishes your presence.

A topic is not a domain with edges. It is how passion focuses itself.

We are building an ever-growing pile of smart leaves that we can organize as we need to at any one moment. Some ways of organizing it—of finding meaning in it—will be grassroots; some will be official. Some will apply to small groups; some will engender large groups; some will subvert established groups. Some will be funny; some will be tragic. But it will be the users who decide what the leaves mean.

The world won't ever stay miscellaneous because we are together making it ours.

CODA: MISC.

Sitting between a brightly lit local ice-cream joint and the unobtrusive Symphony Cleaners, Brookline News and Gift has so thoroughly filled its display windows with odds and ends that none of the interior is visible. Dusty board games are stacked next to a card shuffler. A miniature chrome tank has a watch dial stuck where its entry hatch should be. A genuine Hohner harmonica peeks out from the burled tobacco pipes. By title and display, this is a store that sells *stuff.*

Inside, walkways burrow through the shelves, racks, and displays filling the thousand square feet of the narrow store. As I turn sideways to enter an aisle, my left elbow grazes a "Have a cigar" tote bag, while my right shoulder knocks a small squirt camera off a peg. Nearby are greeting cards, a party mustache, a board game called How to Host a Murder with a photo of the long-dead Vincent Price declaring it "My favorite game," a Sexy Scratcher "lottery" ticket, five-inch pink Cadillac fins suitable for attaching to something, a $14.99 Buddha looking tired and ready to weigh down some paper, "nerd glasses" that will make you the life of the party, umbrellas, a plastic sack of little green army men, meerschaums, brass key rings, and an alphabetized rack of shot glasses with names painted on them. Although there are rough clusters of items—magazines in a rack, digital watches locked in a glass cabinet—the order of the clusters seems random, and every inch around these orderly nests is jammed with stuff that has no obvious place.

When I ask the owner, Michael Wilner, how long he's been here, he answers "Since seven-thirty this morning," a line one suspects he has used before. The store opened in the 1920s, he thinks, and he has been there since June 8, 1963. How does he decide what to stock? "You take a chance." If something sells well—for example, the Sigmund Freud Action Figures—then he orders more. "Once I sold tuna reels like *that,*" he says, snapping his fingers. "I showed people that they sold for twice as much in a catalog. I sold forty or fifty of them. Big tuna reels." Generally, turnover isn't that good, though. Just today he sold two tobacco pipes that he estimates were taking up shelf space for maybe fifteen years. He thinks he still has some greeting cards from when the store opened.

I tell him that Staples has a prototype store and a staff of people who work full-time on arranging items scientifically. "You don't do that?" I ask. He laughs a single "Ha." How does he decide where to put things? "You keep small things around you so they won't get whizzed," he says. "I *think* about putting things together," he adds, but it's clear that that's a plan that gang a-gley more oft than not. I suggest that he probably knows where everything is. "With enough time," he agrees, not sounding confident.

Is Brookline News and Gift the prototype of life in the new world of the miscellaneous? Are we doomed to be looking behind the wax lips, hoping to find the heavy-bristled pipe cleaners but instead starting a cascade of hula girl dashboard statuettes?

While the digital world is far more miscellaneous than any local store could ever be, we don't encounter the miscellaneousness of the digital directly. The worst most of us ever see is a Google list of hits that's gone wrong: We were looking for episodes of *Lost* and instead we got pages about things people can't locate. All we ever see of a third-order miscellany are various orderings of it.

There's order too at Brookline News and Gift. Beyond the physical clusters of board games, tobacco pipes, and magazines, meaning eddies throughout the store:

To Michael Wilner, each turn in the store's organization represents a decision he's made for reasons that were clear at the time.

To one of the regulars, the front of the store may feel suited to his need to choose a cigar, and chat by the counter. The back of the store may be the part for kids and browsers.

To a local schoolkid, everything except the candy counter might be dim and undifferentiated.

To an old-timer, the new items are geological strata simultaneously burying and preserving the past.

To a collector of curios, the dust is metadata.

The truth is that while Brookline News and Gift can have only one arrangement of its stock at any given moment, everyone who steps into it finds some way of taking it. But because Brookline News and Gift is a first-order store, it has only one public way of being organized, determined by which dusty board game is next to which dustier set of vampire teeth.

In the third order, all the ways of organizing a collection can be made public. We can change the visible order to reflect our private meaning. We can share orderings and build on them. Each enhances the meaning of the whole. None has to be given priority. None is more real than another.

So, does the newly miscellanized world look like Michael Wilner's disorderly store?

Yes, but only if we see past its mess to its meaning, for that is what the third order enables.

1. THE NEW ORDER OF ORDER

12 *digital cameras started outselling film cameras:* From Digital Imaging, posted by IT Facts (ZD Net). Available on the Web at http://blogs.zdnet.com/ITFacts/index.php?id=P513.

150 million cell phones with cameras were sold: Study by Info Trends Research Group: "Camera Phone Sales Surge—Report" on Cellular-news.com. Available on the Web at http://www.cellular-news.com/story/10822.shtml.

15 *Library of Congress cataloging:* Interviews with Caroline Arms, Lynn M. El-Hoshy, Rebecca S. Guenther, and Barbara Tillett at the Library of Congress, November 18, 2004.

16 seven million *pages are added to the Web:* Rick Weiss, "On the Web, Research Work Proves Ephemeral: Electronic Archivists Are Playing Catch-Up in Trying to Keep Documents from Landing in History's Dustbin," *Washington Post,* November 24, 2003, p. A08. Available on the Web at http://www.washingtonpost.com/ac2/wp-dyn/A8730-2003Nov23.

19 *Corbis photo cataloging:* Portions of this discussion appeared first in "Taxonomies to Tags: From Trees to Piles of Leaves" in *Release 1.0* (February 2005) and "Point. Shoot. Kiss It Good-Bye" in *Wired* (October 2004). The *Wired* article is available on the Web at http://www.wired.com/wired/archive/12.10/photo.html.

21 *Flickr photo sharing:* Numbers accurate as of August 2006. From an email from Stewart Butterfield, August 29, 2006.

2. ALPHABETIZATION AND ITS DISCONTENTS

24 *Volapük:* Paul LaFarge, "Pük, Memory: Why I Learned a Universal Language No One Speaks," *Village Voice,* August 2–8, 2000. Available on the

Web at http://www.villagevoice.com/arts/0031,lafarge,16942,12.html. See also "Volapük" at Omniglot: A Guide to Written Language. Available on the Web at http://www.omniglot.com/writing/volapuk.htm.

24 *Charles Luthy's Universal Alphabet:* Chas. T. Luthy, *The Universal Alphabet* (Self-published, 1918).

26 *Early days of alphabetization:* Lloyd W. Daly, *Contributions to a History of Alphabetization in Antiquity and the Middle Ages* (Brussels: Latomus, 1967).

27 *"Nowe if the word, which thou art desirous":* Robert Cawdrey, "To the Reader," in *A Table Alphabetical*, ed. Raymond Siemens (1994). Available on the Web at http://www.library.utoronto.ca/utel/ret/cawdrey_reader.html.

papyrus tax rolls: Daly, *Contributions to a History of Alphabetization.*

the great French Encyclopédie: Philipp Blom, *Enlightening the World* (New York: Palgrave, 2004).

28 *Mortimer Adler:* Mortimer J. Adler, *A Guidebook to Learning: For a Lifelong Pursuit of Wisdom* (New York: Macmillan, 1986).

"By the bye, what a strange abuse": Samuel Taylor Coleridge to Thomas Wedgwood, Letter 116, February 10, 1803, in *Biographia Epistolaris,* vol. 1, ed. Arthur Turnbull (London: G. Bell and Sons, 1911). Available on the Web at http://www.fullbooks.com/Biographia-Epistolaris-Volume-16.html.

29 *"your Presbyterian bookmakers":* Adler says those bookmakers "must have been" the creators of the *Encyclopaedia Britannica.* Adler, *Guidebook,* p. 17.

Encyclopedia Metropolitana: "Encyclopedia," *Encyclopaedia Britannica* online.

Adler biographical material: "Dr. Adler's Biography," The Radical Academy, available on the Web at http://radicalacademy.com/adlerbio.htm.

"1,690 ideas found to be respectable": Dwight Macdonald, "The Book-of-the-Millennium Club," *New Yorker,* November, 29, 1952. Available on the Web at http://www.writing.upenn.edu/~afilreis/50s/macdonald-great-books.html.

cost half of the $2 million: William Benton, introduction to Mortimer J. Adler, *Great Ideas from the Great Books* (New York: Washington Square Press, 1961), p. vii.

30 *hopelessly outdated:* Macdonald, "The Book-of-the-Millennium Club."

31 *less than 3 percent change:* From a note Adler posted to the Western Canon Mailing List in September 1997. Available on the Web at http://books.mirror.org/gb.sel1990.html.

"We do not scorn chronology or alphabetization": Joseph J. Esposito, "Redesigning, Not Reinventing, Encyclopaedia Britannica," *Proceedings of the Third Symposium on Gateways, Gatekeepers, and Roles in the Informa-*

tion Omniverse (Washington, D.C.: Association of Research Libraries, 1994).

32 *natural "joints":* In the *Phaedrus,* Plato writes, "The second principle is that of division into species according to the natural formation, where the joint is, not breaking any part as a bad carver might" (265c, in a translation by Benjamin Jowett, available on the Web at http://ccat.sas.upenn.edu/jod/texts/phaedrus.html).

The philosopher Ian Hacking: Ian Hacking, *The Social Construction of What?* (Cambridge: Harvard University Press, 1999).

South African apartheid: Geoffrey C. Bowker and Susan Leigh Star, *Sorting Things Out: Classification and Its Consequences* (Cambridge: MIT Press, 1999), chapter 6.

33 *"GayPA":* Lynne Lamberg, "Gay Is Okay with the APA," *Medical News & Perspectives,* August 12, 1998. Available on the Web at http://www.soulforce.org/main/psychiatric.shtml. Also Alix Spiegel, "81 Words," *This American Life,* Episode 204, January 18, 2002, and Matt and Andrej Koymasky, "Dr. John E. Fryer, MD," *The Living Room,* June 16, 2004, http://andrejkoymasky.com/liv/fam/biof2/frye3.html.

If the syndrome isn't in the DSM: Bowker and Star, *Sorting Things Out,* p. 46.

"sociopathic personality disturbance": Joseph Nicolosi, "The Removal of Homosexuality from the Psychiatric Manual." Available on the Web at http://www.catholicsocialscientists.org/Symposium2–Nicolosi–mss.htm.

Ronald Gold: Ronald Gold, "Three Stories of Gay Liberation," *Queer Stories,* 2004. Available on the Web at http://www.queerstories.org/custom.html.

Gold described: Michelangelo Signorile, "Our Gay Century," *Advocate,* January 18, 2000. Available on the Web at http://www.signorile.com/articles/advogc.html.

The speech led him: Richard Lafferty, "Queerly Ill: The Rise and Fall of the Illness of Homosexuality," October 20, 2000. Available on the Web at http://www.lafferty.ca/writing/queerly_ill/lafferty_queerly_ill.pdf.

"the resolution that a year later": Gold, "Three Stories of Gay Liberation."

homosexuality was considered a problem only: Nicolosi, "The Removal of Homosexuality from the Psychiatric Manual."

34 *"If our hearts were as pure, as chaste":* John Milton, "On the Music of the Spheres." Available on the Web at http://www.geocities.com/iconostar/milton-spheres.htm.

Pythagoras therefore figured: To be precise, it was the distance between the celestial spheres that contained the planets.

35 *the Great Chain of Being:* Arthur O. Lovejoy, *The Great Chain of Being: The Study of the History of an Idea* (Cambridge: Harvard University Press, 1936).

35 *The controversy had grown:* Andrew Bridges, "Massive Body Discovered Beyond Pluto," AP Online; October 7, 2002.

36 *Brown reasoned:* Michael McCarthy, "Meet Xena, the Tenth Planet in the Solar System," *Independent* (London, Eng.), August 1, 2005.

"When a Caltech astronomer, Michael Brown": Editorial, "Too Many Planets Numb the Mind," August 2, 2005. (Obviously, there's some disagreement about the size of Xena. The *Independent* reported that it was in fact larger than Pluto.)

37 *Bode's law:* Karl Kruszelnicki, "Planetary Alignment—Part 1," *Great Moments in Science,* ABC News (Australia). For the mathematically inclined, this article explains the flaw in Bode's law: John N. Harris, "Spira Solaris: Archytas Mirabilis," Part 1, November 16, 2001. Available on the Web at http://www.spirasolaris.ca/sbb4a.html.

formal definition of a planet: Robert Roy Britt, "Defining 'Planet': Newfound World Forces Action," MSNBC, August 2, 2005. Available on the Web at http://msnbc.msn.com/id/8800646/. See also Jeff Hecht, "Tenth Planet Discovered in Outer Solar System," July 30, 2005, NewScientist .com news service. Available on the Web at http://www.newscientist space.com/article.ns?id=dn7763.

"a group that thinks planet *is a cultural term":* Alan Stern, phone interview, September 15, 2005.

Stern's definition of a planet: Alan S. Stern, "Gravity Rules," *Space Review* online, March 29, 2004. Available on the Web at http://www.thespace review.com/article/123/1.

38 *"Schoolkids can't name all the mountains":* Stern, interview.

2009 meeting: Robert Roy Britt, "Pluto: Down but Maybe Not Out," Space.com, August 31, 2006. Available on the Web at http://www.space .com/scienceastronomy/060831_planet_definition.html.

adjectives to the planets: Robert Roy Britt, "Definition Debate: Planets May Soon Get Adjectives," Space.com, September 21, 2005. Available on the Web at http://www.space.com/scienceastronomy/050921_planet _definition.html.

40 *triads of elements:* Eric R. Scerri, "The Evolution of the Periodic System," *Scientific American,* vol. 279, issue 3, September 1998.

41 *"The properties of the elements":* Linus Pauling, "Periodic Law and Table," Britannica Guide to the Nobel Prizes, 1997. Available on the Web at http://www.britannica.com/nobel/macro/5001_20_45.html. See also "Alexandre-Émile Béguyer de Chancourtois" at the Royal Society of Chemistry's Web site: http://www.chemsoc.org/networks/learnnet /periodictable/pre16/develop/chancourtois.htm.

John Newlands and the Chemical Society: Scerri, "Evolution of the Periodic System."

41 *Dmitrii Ivanovich Mendeleev:* Michael D. Gordin, *A Well-Ordered Thing: Dmitrii Mendeleev and the Shadow of the Periodic Table* (New York: Basic Books, 2004), p. 26.

He also did not believe that matter: Nathan M. Brooks, "Developing the Periodic Law: Mendeleev's work during 1869–1871," *Foundations of Chemistry,* vol. 4 (Dordrecht, Netherlands: Kluwer Academic Publishers, 2002), pp. 127–47.

a version of solitaire: "Chemistry: The Periodic Table of the Elements," *The New York Public Library Desk Reference* (Carlsbad, Calif.: Compton's NewMedia, 1995).

42 *The alternatives include:* Emil Zmaczynski, electron shell table, http://library.thinkquest.org/C0110203/othertables.htm; Ed Perley, "A Circular Periodic Table of the Elements," http://www.nfinity.com/~exile/periodic.htm; "Electric Prism Technology, Periodic Spiral," http://www.periodicspiral.com/; and http://www.chemicalgalaxy.co.uk/. See also "A Galaxy of Elements," PhysOrg.com, August 15, 2005, http://www.physorg.com/news5816.html. ChemistryCoach has many more examples on the Web at http://www.chemistrycoach.com/periodic_tables.htm.

43 *Philip Stewart:* Yes, "plant," not "planet."

Railsback's chart: Alexandra Goho, "The Nature of Things: Attempts to Change the Periodic Table Raise Eyebrows," *Science News,* October 25, 2003.

45 *As Umberto Eco says:* Umberto Eco, *Kant and the Platypus: Essays on Language and Cognition,* trans. Alastair McEwen (San Diego: Harcourt, 1997), p. 53.

3. THE GEOGRAPHY OF KNOWLEDGE

46 *Dewey decimal system:* Wayne A. Wiegand, "The 'Amherst Method': The Origins of the Dewey Decimal Classification Scheme," *Libraries and Culture,* vol. 33, no. 2 (Spring 1998), p. 175. Available on the Web at http://www.gslis.utexas.edu/~landc/fulltext/LandC_33_2_Wiegand.pdf. Only 25 percent of U.S. college libraries, mainly small ones, use the Dewey system.

48 *Melvil Dewey biography:* Wayne A. Wiegand, *Irrepressible Reformer: A Biography of Melvil Dewey* (Chicago: American Library Association, 1996). Wiegand's excellent biography is the source of much of the information in this chapter about Dewey's life.

50 *scrolls were listed alphabetically:* "Library," *Encarta* online, available on the Web at http://encarta.msn.com/encyclopedia_761564555_16/Library_(institution).html.

Gutenberg: Tjalda Nauta, "From Gutenberg to the Gutenberg Project—Lost in the Information Glut," *Issues in Teaching and Learning,* vol. 2,

Rhode Island College. Available on the Web at http://www.ric.edu/itl/issue02/articleNauta.html.

50 *Panizzi's great offense:* Brendan A. Rapple, "Coping with Catalogues: Thomas Carlyle in the British Museum," *Contemporary Review,* December 1, 1996.

A death sentence awaited: Tom Rosenthal, "Marx's Office," *New Statesman,* January 8, 2001.

51 *"The fat pedant":* Rapple, "Coping with Catalogues."

Panizzi's rules: "Institutional History Division," Smithsonian Institution Archives, available on the Web at http://siarchives.si.edu/sia/fy99/ihd.htm.

Charles Coffin Jewett's card catalog: Suzanne C. Pilsk et al., "Organizing Corporate Knowledge: The Ever-Changing Role of Cataloging and Classification," *Information Outlook,* April 2002. Available on the Web at http://www.findarticles.com/p/articles/mi_m0FWE/is_4_6/ai_95200282.

Jewett came up with the idea: "Charles Coffin Jewett (1816–1868)," *The Hutchinson Encyclopedia of Science,* September 22, 2003.

"Of this I am inclined to be a friend": Cited in Wiegand, "The 'Amherst Method.'"

52 *Georg Wilhelm Friedrich Hegel:* G. W. F. Hegel, "Section One: Modern Philosophy in Its First Statement, A: Bacon," *Hegel's Lectures on the History of Philosophy.* Available on the Web at http://www.marxists.org/reference/archive/hegel/works/hp/hpbacon.htm.

53 *Dewey's third big idea:* "The 'Amherst Method,' Wiegand," p. 180.

55 *In the 1980s:* Interview with Joan S. Mitchell, editor in chief of the Dewey Decimal Classification, Washington, D.C., November 18, 2004.

The 000 class was renamed: Joan S. Mitchell, "DDC 22 and Beyond: Dewey in a Global and Local Context," presentation to the Australian Committee on Cataloguing, Sydney, Australia, July 25, 2003.

4. LUMPS AND SPLITS

65 *"(a) belonging to the emperor":* Jorge Louis Borges, "The Analytical Language of John Wilkins," ed. Jan Frederik Solem et al., trans. Lilia Graciela Vázquez. Available on the Web at http://www.alamut.com/subj/artiface/language/johnWilkins.html.

66 *too broad a category:* As the French philosopher Michel Foucault says about Borges's list: "The common ground on which such meetings are possible has itself been destroyed." Foucault is perhaps our age's deepest thinker about the history and significance of taxonomy. Michel Foucault, *The Order of Things* (New York: Random House, 1970), p. xvi.

67 *the earliest map:* "Time Charts of Cartography," on the Web at http://www.henry-davis.com/MAPS/Ancient%20Web%20Pages/100mono.html. For

a good illustration of the Catal Hyük map, see Michael Lahanas, "Ptolemy's Geography and Maps," on the Web at http://www.mlahanas.de/Greeks/PtolemyMap.htm. For a photo and interpretation of the Ga-Sur map, see the Henry Davis Consulting site: http://www.henry-davis.com/MAPS/Ancient%20Web%20Pages/100D.html.

69 *"To say that the ideas are patterns":* Aristotle, *Metaphysics,* Alpha, l. 991a20, 991b, 991b26.

Aristotle said that a category: I'm not using the word *category* in Aristotle's technical sense. For him, the categories correspond to the ten basic questions one can ask about something: What is it made of?, What is it doing?, and so on.

This lumping and splitting: Aristotle, *Metaphysics,* Alpha, l. Zeta, 1030a.

71 *"A lumper takes things":* Interview with Seth Maislin, February 2004.

72 *Tournefort introduced the notion:* Norbert Ross, "Evolution and Devolution of Knowledge: A Tale of Two Biologies," *Journal of the Royal Anthropological Institute,* June 1, 2004.

That pared down the number: Michel Foucault, *The Order of Things,* pp. 77, 134, 141.

73 *Because the system utilized:* David Knight, *Ordering the World: A History of Classifying Man* (London: Burnett, 1981), p. 77. This is a thoughtful and readable history of ideas.

knew Genesis by heart: Lisbet Koerner, *Linnaeus: Nature and Nation* (Cambridge: Harvard University Press, 1999).

"binomial" system: Heinz Goerke, *Linnaeus* (New York: Scribner's, 1973), pp. 100–101.

74 *Twice a week:* Koerner, *Linnaeus,* p. 41.

one German botanist: Ibid., p. 27. The botanist was Johann Georg Siegesbeck.

75 *Because the specimens are made of atoms:* Foucault says that collections become "the documents of this new history." *Order of Things,* p. 131.

76 Homo sylvestris: David M. Knight, email correspondence, September 2006.

ranking each of the species: "Linnaeus had it constantly in mind: 'The closer we get to know the creatures around us, the clearer is the understanding we obtain of the chain of nature, and its harmony and system, according to which all things appear to have been created.' " Sten Lindroth, "The Two Faces of Linnaeus," in Tor Frangsmyr, ed., *Linnaeus: The Man and His Work* (New York: Science History Publications, 1994), p. 16.

77 *"depicted in maps or paintings":* Koerner, *Linnaeus,* p. 40.

IBM Business Consulting Services: Portions of this section appeared first in *Forrester Research* magazine, April 2005, and *Release 1.0,* February 2005.

79 *S. R. Ranganathan's biography and career:* See the biography of Ranganathan by his son, Yogeshwar Ranganathan, *S. R. Ranganathan: Pragmatic Philosopher of Information Science: A Personal Biography* (Mumbai: Bhavan's Book University, 2000).

80 *It is not a simple:* Elaine Svenonius, *The Intellectual Foundation of Information Organization* (Cambridge: MIT Press, 2000), pp. 175–76. Robert J. Clushko explains the notation in his class notes for IS 202, University of California, Berkeley, September 20, 2005. (He refers to fifty-five instead of forty-four, presumably a typo.) Available on the Web at http://www2.sims.berkeley.edu/courses/is202/f05/LectureNotes/202-20050920.pdf.

81 *"sees beyond the phenomenal":* S. R. Ranganathan, "Study Circle and Joy," *Library Herald,* vol. 7, no. 4 (January 1965), pp. 217–18, cited in M. P. Satija, *S. R. Ranganathan and the Method of Science* (Delhi: Aditya Prakashan, 1992), p. 53.

5. THE LAWS OF THE JUNGLE

86 *"Alison Lukes et Cie":* On the Web at http://www.alisonlukes.com/alison.html.

Pepsico says that about an eighth: Carol Hymowitz, "The New Diversity," *Wall Street Journal,* November 14, 2005, p. R1.

When he was editing Jarhead: In an interview on NPR's *All Things Considered,* November 8, 2005.

the "Slob Sisters": Pam Young and Peggy Jones, *Sidetracked Home Executives* (New York: Warner Books, 2001), p. 7.

87 *we should be suspicious:* Stephen Jay Gould, "A Tree Grows in Paris," *The Lying Stones of Marrakech: Penultimate Reflections in Natural History* (New York: Harmony Books, 2000), pp. 115–43.

fifteen thousand species: Carol Kaesuk Yoon, "In the Classification Kingdom, Only the Fittest Survive," *New York Times,* October 11, 2005.

Linnaeus loved plants: David Knight, *Ordering the World: A History of Classifying Man* (London: Burnett Books, 1981). Knight, in email correspondence (November 2005), speculated about why Linnaeus did such a poor job differentiating categories within *Vermes.* Knight wrote that Linnaeus had an affinity for plants: "his house near Uppsala papered with plant pictures, his reputation made in the Netherlands with botany, and the 'disciples' and their dissertations being concerned mostly with plants." Further, Knight says, Lamarck was a physician and botany was considered to have much more to teach doctors than did the study of worms. Knight concludes that when it came to worms, how they were generated was ambiguous and their taxonomy was "tricky," so "they were best left to future scholars."

88 *"a kind of chaos":* Gould, *Lying Stones,* p. 130.

89 *"NewsCodes":* The IPTC home page on the Web is http://www.iptc.org/NewsCodes/nc_ts-table01.php?TsByName=iptc-subjectcode. (Thanks to Kurt Starsinic for bringing this to my attention.)

Getty Art and Architecture Thesaurus: Joseph Busch, telephone interview, March 2004.

92 *Delicious's creator:* Joshua Schachter, several interviews, beginning in January 2005.

94 *225 million photos onto Flickr:* Email from Stewart Butterfield, Flickr's cofounder, August 29, 2006. Butterfield says that if you count "tags that have been used by at least five different people for a total of forty instances," in order to remove completely idiosyncratic tags and misspellings, the number of unique tags goes down to 201,839.

96 *The system the BBC:* Interview with Sarah Hayes, July 2005.

98 *In fact, a Wikipedia article isn't:* There's information about Wikipedia's servers at http://meta.wikimedia.org/wiki/Wikimedia_servers. Much of the information in this section came from Brion Vibber and Tim Starling on a chat on the Wikipedia technical IRC channel on November 1, 2005.

103 *Public Library of Science:* Discussion with Hemai Parthasarathy, October 20, 2006. Blogged at http:// www.hyperorg.com/blogger/mtarchive/berkman_plos_open_access_scien.html.

6. SMART LEAVES

107 *"It could cause eye damage":* Tony Seideman, "The History of Barcodes," *American Heritage of Invention and Technology,* vol. 10 (Fall 1994), pp. 24–31. Available on the Web at http://www.basics.ie/History.htm. More information on bar-code history can be found in Russ Adams's "Bar Code History," on the Web at http://www.adams1.com/pub/russadam/history.html, and in George J. Laurer's "Development of the U.P.C. Symbol," October 2001, on the Web at http://bellsouthpwp.net/1/a/laurergj/UPC/upc_work.html.

108 *Today there are about five billion:* From the Uniform Code Council (now GS1 US), on the Web at http://www.uc-council.org/upc_background.html.

PULP: Liz Lawley, "What I've Been Working On," Mamamusings blog, June 27, 2006, on the Web at http://mamamusings.net/archives/2006/06/27/what_ive_been_working_on.php; I also attended Lawley's session on this at the Foo Camp conference in August 2006.

109 *UPC code adoption:* Rob Baker, "UPC Bar Code Session Draws a Hot Following," *WWD,* January 13, 1987; Carl Barbati, "UPOC Council Acting

on Faulty Bar Codes," *Supermarket News,* August 13, 1984; Rob Baker, "UPC Wins Top Endorsement," *WWD,* June 26, 1986; Rob Baker, "Uniform Code Council Approves Seafood UPC," *Marine Fisheries Review,* January 1, 1990; Elliot Zwiebach, "Sept. 18 Parley Set to Mull Longer UPC," *Supermarket News,* September 8, 1986. See also Holly A. Cobb, "OK Random-wt. 24-digit UPC," *Supermarket News,* March 30, 1987.

109 *In a 1986 study:* Information in this paragraph comes from Rachel Spevack, "UPC Bar Codes' Time Has Come, Bullock's V-P Informs Retailers," *Daily News Record,* January 13, 1987, and Russ Adams, "Barcode FAQ," on the Web at http://www.adams1.com/pub/russadam/faq.html.

UPC codes consist of thirteen: EAN: Warren Hagey, "Understanding Bar Codes," 1998, on the Web at http://educ.queensu.ca/~compsci/units/encoding/barcodes/undrstnd.html.

110 *UNSPSC:* You can download the entire code at the UNSPSC Web site, http://www.unspsc.org. There's also a white paper available on the site: "Using the UNSPSC: Why Coding and Classifying Products Is Critical to Success in Electronic Commerce," September 1998, updated October 2001.

111 *RFID uses:* "to track cows" in "General RFID Information," *RFID Journal,* on the Web at http://www.rfidjournal.com/faq/16/56; "to detect U.S. Energy Department prohibited materials" in Florence Olsen, "Feds Find RFID Uses," *Federal Computer Week,* May 31, 2005, on the Web at http://www.fcw.com/article89026-05-31-05-Web; "to track all of the cargo and equipment used in the Iraq War": David C. Wyld, *RFID: The Right Frequency for Government* (IBM Center for the Business of Government, 2005), p. 36, on the Web at http://www.businessofgovernment.org/pdfs/WyldReport4.pdf.

Kroger estimates: Mary Catherine O'Connor, "Kroger Turning to RFID to Stay Fresh," *RFID Journal,* December 20, 2005, on the Web at http://www.rfidjournal.com/article/articleview/2055/1/1/.

A University of Arkansas study: Laurie Sullivan, "Wal-Mart RFID Trial Shows 16% Reduction in Product Stock-Outs," *InformationWeek,* October 14, 2005. Available on the Web at http://informationweek.com/story/showArticle.jhtml?articleID=172301246.

Three Virginia hospitals: Jonathan Collins, "Hospitals Get Healthy Dose of RFID," *RFID Journal,* April 27, 2004. Available on the Web at http://www.rfidjournal.com/article/view/920.

become "spime": Bruce Sterling, "When Blobjects Rule the Earth," transcript of keynote speech at Special Interest Group on Graphics and Interactive Techniques (SIGGRAPH) Conference in Los Angeles, Calif., August 8–12, 2004. Available on the Web at http://www.boingboing.net/images/blobjects.htm.

111 *In addition to the BBC's gargantuan library:* Tom Coates's presentation at O'Reilly Emerging Technology Conference, April 2005, Santa Barbara, Calif. Thanks to Tom for sending me his presentation.
shepherded by John Good and Carol Owens: Interview with John Good, October 14, 2004.

112 *"how people find programming":* Interview with Tom Coates, August 2005.
So their system automatically creates: Tom Coates left the BBC in the fall of 2005 and reports that the project has slowed down.

113 The Seafood List: U. S. Food and Drug Administration, Center for Food Safety and Applied Nutrition. The "Search the Seafood List" page is available at http://vm.cfsan.fda.gov/~frf/seaintro.html.

114 *Information about the All Species Foundation and ZooBank:* Carol Kaesuk Yoon, "In the Classification Kingdom, Only the Fittest Survive," *New York Times,* October 11, 2005.

115 *LSID project:* Salvatore Salamone, "LSID: An Informatics Lifesaver," *Bio-IT World.com,* December 16, 2005. Available on the Web at http://www.bio-itworld.com/archive/011204/lsid.html. See also Lee Belbin, "An ID Tag for Biodiversity Information Objects," Global Biodiversity Information Facility, June 27, 2006, on the Web at http://www.gbif.org/Stories/STORY1143196078.
The Tree of Life Web project: On the Web at http://tolweb.org/tree/phylogeny.html.

116 *Birders in southwestern Africa:* C. Cohen and C. N. Spottiswoode, *Essential Birding—Western South Africa* (Cape Town: Struik Publishers, 2000). Adapted on the Web at http://birdingafrica.maxitec.co.za/birdingafrica/Resources_Taxonomy.html. For technical details on classifying African larks: Peter G. Ryan and Paulette Bloomer, "Long-Billed Lark Complex: A Species Mosaic in Southwestern Africa," *Auk,* January 1999. Available on the Web at http://www.findarticles.com/p/articles/mi_qa3793/is_199901/ai_n8836359/pg_1.
essentialism—*the idea that:* For a good discussion of Aristotle on species and essence, see "Aristotle's Metaphysics" in *Stanford Encyclopedia of Philosophy,* October 8, 2000, revised November 7, 2003, on the Web at http://plato.stanford.edu/entries/aristotle-metaphysics/.David Hull argues convincingly that Aristotle's views were subtler and more complex. For example, he thought some species entirely died out during the winter and thus were not constantly present. David L. Hull, "Linné as an Aristotelian," in *Contemporary Perspectives on Linnaeus,* ed. John Weinstock (Lanham, Md.: University Press of America, 1985).

117 *"It is really laughable":* Francis Darwin, ed., *The Life and Letters of Charles Darwin, Including an Autobiographical Chapter,* vol. 2 (London: John

Murray, 1877), p. 88. The passage is dated December 24, 1856. Cited in "Species," *Stanford Encyclopedia of Philosophy*, July 4, 2002. Available on the Web at http://plato.stanford.edu/entries/species/#DoeSpeCatExi.

117 *"we shall have to treat species":* Charles Darwin, *Origin of Species*, p. 485.
One expert: Marc Ereshefsky, "Species and the Linnean Hierarchy," in *Species: New Interdisciplinary Essays*, ed. Robert A. Wilson (Cambridge: MIT Press, 1999), p. 290.
"Classes, orders, genera": In his correspondence with John Manners in 1814.
As the philosopher: James Danaher, "The Fallacy of the Single Real Essence," *Philosopher*, vol. 88, no. 2. Available on the Web at http://www.the-philosopher.co.uk/esse.htm.

119 *Melvil Dewey himself:* Wayne A. Wiegand, *Irrepressible Reformer: A Biography of Melvil Dewey* (Chicago: American Library Association, 1996), p. 54.
you can sometimes tell if a card: John Seely Brown and Paul Duguid, *The Social Life of Information* (Cambridge: Harvard Business School Press, 2000).

120 *LibraryLookup:* Available on the Web at http://weblog.infoworld.com/udell/stories/2002/12/11/librarylookup.html.
Harvard's experimental H2O site: On the Web at http://h2obeta.law.harvard.edu/home.do.

121 *the official First Folio edition:* The folio has been scanned and is available on the Web at http://ise.uvic.ca/Library/plays/Ham.html.
"There is a world of difference": James Shapiro, *A Year in the Life of William Shakespeare: 1599* (New York: HarperCollins, 2005), p. 306.
Large publishers buy: Interview with Carol Cooper, senior director, Standards Services, Bowker, November 2005.

122 *Hickey's project, xISBN:* Telephone interview with Tom Hickey, OCLC, October 2005.
Functional Requirements for Bibliographic Records (FRBR) standard: IFLA Study Group on the Functional Requirements for Bibliographic Records, *Functional Requirements for Bibliographic Record: Final Report*, UBCIM Publications, New Series, vol. 19 (The Hague: International Federation of Library Associations and Institutions, 1998). Available on the Web at http://www.ifla.org/VII/s13/frbr/frbr.pdf.

125 *Microsoft Research AURA project:* The project's beta site is available on the Web at http://aura.research.microsoft.com/Aura/DesktopDefault.aspx?tabName=Home.

126 *When he scanned his favorite breakfast food:* Eric Bender, "Social Lives of a Cell Phone," *Technology Review*, July 12, 2004. Available on the Web at http://nasw.org/users/Bender/social_cell_phones.html.

126 *The DNA bar-coding project:* The Canadian Centre for DNA Barcoding Web site is http://www.barcodinglife.ca.

7. SOCIAL KNOWING

129 *Digg.com:* "About Digg," June 28, 2006, on the Web at http://digg.com/about.

130 *"narrow down your results":* Email interview with Kevin Burton, January 13, 2006. This feature was planned for release at the end of January 2006.

131 The Daily Me: Nicholas Negroponte, *Being Digital* (New York: Alfred A. Knopf, 1995).

"wisdom of crowds": James Surowiecki, *The Wisdom of Crowds: Why the Many Are Smarter Than the Few and How Collective Wisdom Shapes Business, Economies, Societies, and Nations* (New York: Doubleday, 2004).

"A blog is a species": Michael Gorman, "Revenge of the Blog People!," LibraryJournal.com, February 15, 2005. Available on the Web at http://www.libraryjournal.com/article/CA502009.html?display=BackTalkNews&industry=BackTalk&industryid=3767&verticalid=151.

Some librarians . . . were outraged: Sarah Houghton, "Michael Gorman Is Irresponsible," Librarian in Black blog, February 25, 2005, on the Web at http://librarianinblack.typepad.com/librarianinblack/2005/02/michael_gorman_.html; Karen Schneider, "Gorman on Bloggers," Free Range Librarian blog, February 24, 2005, on the Web at http://freerangelibrarian.com/archives/022405/gorman_on_bloggers.php; "Turkey ALA King," BatesLine blog, February 27, 2005, on the Web at http://www.batesline.com/archives/001387.html.

132 *"a lovely idea":* Simon Waldman, "Who Knows?" *Guardian Unlimited,* October 26, 2004. Available on the Web at http://technology.guardian.co.uk/online/news/0,12597,1335892,00.html.

"The user who visits": Robert McHenry, "The Faith-Based Encyclopedia," *Tech Central Station,* November 15, 2004. Available on the Web at http://www.techcentralstation.com/111504A.html.

133 *To protect its trademark:* "OCLC Sues New York Library–Themed Hotel," LibraryJournal.com, September 29, 2003. Available on the Web at http://www.libraryjournal.com/index.asp?layout=article&articleid=CA325514&publication=libraryjournal.

NAR protects its members' interests: On the clout of the NAR, see Glen Justice, "Lobbying to Sell Your House," *New York Times,* January 12, 2006.

134 *"Nobel laureates and Pulitzer Prize winners":* "The *Encyclopaedia Britannica* Board of Advisors," on the Web at http://corporate.*Britannica*.com/board/.

"Seigenthaler Affair": "John Seigenthaler Sr.," Wikipedia, on the Web at

http://en.wikipedia.org/wiki/John_Seigenthaler_Sr. Yes, I am aware of the irony.

135 *"At age 78":* John Seigenthaler, "A False Wikipedia 'Biography,'" *USA Today,* November 29, 2005. Available on the Web at http://www.usatoday.com/news/opinion/editorials/2005-11-29-wikipedia-edit_x.htm.

As Jimmy Wales: In a private meeting. Wales said something similar in his keynote at the Wikimania conference in Cambridge, Massachusetts, in August 2006.

136 *The personal peccadilloes of the greatest contributors:* Simon Winchester, *The Professor and the Madman: A Tale of Murder, Insanity, and the Making of the* Oxford English Dictionary (New York: HarperCollins, 1998).

137 *Swift Boat Veterans for Truth:* "Swift Vets and POWs for Truth," as of January 23, 2006, on the Web at http://en.wikipedia.org/wiki/Swift_Vets_and_POWs_for_Truth.

138 *On rare occasions:* Stacy Schiff, "Know It All: Can Wikipedia Conquer Expertise?" *New Yorker,* July 31, 2006.

The vote also decided: "Talk:Gdansk/Vote," on the Web at http://en.wikipedia.org/wiki/Talk:Gdansk/Vote.

(Imagine if we could): I wish I could remember who pointed this out to me and used this example.

Wikipedia works as well as it does: Jim Giles, "Internet Encyclopaedias Go Head to Head," *Nature,* December 15, 2005, pp. 900–901. Available on the Web at http://www.nature.com/nature/journal/v438/n7070/full/438900a.html.

Bomis: From Brian Lamb's interview with Wales on C-SPAN, September 2005, cited in Jimmy Wales's entry at Wikipedia. Wales disputes that Bomis found soft-core porn; he says it found R-rated movies.

the review process at Nupedia: "Nupedia," Wikipedia, on the Web at http://en.wikipedia.org/wiki/Nupedia.

139 *half of the edits:* From a presentation by Wales at *Nature* magazine headquarters in December 2005. Also reported from Wales's presentation at the Reboot conference, June 2005. See http://reboot.dk/reboot7/show/Participants+Notes+from+The+Intelligence+of+Wikipedia. See also http://www.hyperorg.com/blogger/mtarchive/wikipedias_long_tail.html.

Far from hiding this hierarchy: Aaron Swartz, "Who Writes Wikipedia?" September 4, 2006, on the Web at http://www.aaronsw.com/weblog/whowriteswikipedia.

140 *Wikipedia has an arsenal:* The list of notices is on the Web at http://en.wikipedia.org/wiki/Wikipedia:Template_messages.

141 *When* Nature *magazine released:* "Wikipedia: External peer review/Nature December 2005/Errors," on the Web at http://en.wikipedia.org/wiki/Wikipedia:External_peer_review/Nature_December_2005/Errors.

141 *Wikipedia even has a page listing errors:* "Wikipedia: Errors in the Encyclopaedia Britannica that have been corrected in Wikipedia," on the Web at http://en.wikipedia.org/wiki/Wikipedia:Errors_in_the_Encclop%C3%A6dia_Britannica_that_have_been_corrected_in_Wikipedia.

145 *The CIO of the investment bank: Business Week,* November 28, 2005.

Suzanne Stein of Nokia's Insight & Foresight: "Nokia Uses Socialtext as an Alternative to Email" (marketing write-up), on the Web at http://www.socialtext.com/node/40.

Michelin China: Jean Noel Simonnet, "The Wiki Success Story of Michelin China" (marketing write-up), December 2, 2004, on the Web at http://twiki.org/cgi-bin/view/Main/TWikiSuccessStoryOfMichelinChina.

146 *Disney, SAP, and some major pharmaceutical companies:* Rob Hof, "Do-It-Yourself Software for All?" *Business Week,* October 6, 2004. On the Web at http://www.businessweek.com/technology/content/oct2004/tc20041106_2351.htm.

With over 50 million known blogs: Figures from Technorati, as reported in an email message by Kevin Marks, September 15, 2006.

8. WHAT NOTHING SAYS

148 *Some labels are so dumb they're famous:* Some of these came from eBaum's World, on the Web at http://www.ebaumsworld.com/labels.shtml. There are no accreditations or references, so Warning: Some of the labels cited may be apocryphal.

150 *Automobile Club of Southern California:* "Road to the Past Marked by Signs: Auto Club's Vintage Markers Are Featured in a Visalia Display," *Fresno Bee,* May 7, 2000.

Manual on Uniform Traffic Control Devices: The manual is on the Web at http://mutcd.fhwa.dot.gov/.

151 *"WalMart online shoppers":* Post at Digg.com on the Web at http://digg.com/movies/WalMart_website_compares_African_American_leaders_to_Apes. I've corrected some obvious typos that got in the way of reading.

152 *Amazon said that its software:* Laurie J. Flynn, "Amazon Amends Abortion Search Results," *International Herald Tribune,* March 20, 2006. Available on the Web at http://www.iht.com/articles/2006/03/20/business/amazon.php.

153 *Shakespeare was trying to sound old-fashioned:* James Shapiro, *A Year in the Life of Shakespeare: 1599* (New York: HarperCollins, 2005), pp. 290–91.

154 *"Bumptunes":* "Bumptunes Released," *iPod Hacks,* March 7, 2005, on the Web at http://ipodhacks.com/article.php?thold=-1&mode=flat&order=0&sid=1272.

156 *Maps "lie on purpose":* Interview with Howard Veregin, January 2004.

157 *Google Maps spurred quick-witted:* Apartment listings on the Web at http://

www.indiesoft.com/craigsmaps/; http://paulrademacher.com/housing/; Google Maps in Flickr on the Web at http://userscripts.org/scripts/show/1574; prison in Tunisia on the Web at http://kitab.nl/tunisianprisoners map.

158 *Jefferson's Koran:* Kevin J. Hayes, "How Thomas Jefferson Read the Qur'an," *Early American Literature,* March 22, 2004. As evidence that Jefferson took pains with his catalog, Hayes cites a letter Jefferson wrote after he sold his collection to the Library of Congress: "The form of the catalogue has been much injured in the publication; for although they have preserved my division into chapters, they have reduced the books in each chapter to alphabetical order, instead of the chronological or analytical arrangements I had given them."

159 *"an important means of self-expression":* Howard Parnell, "Downloading Empathy to Your iPod," *Washington Post,* March 1, 2006. Available on the Web at http://www.washingtonpost.com/wp-dyn/content/article/2006/03/01/AR2006030100635.html.

163 *"Most users fear the presence":* danah boyd, "Friendster and Publicly Articulated Social Networking," *Conference on Human Factors and Computing Systems* (Vienna: ACM, April 24–29, 2004). Available on the Web at http://www.danah.org/papers/CHI2004Friendster.pdf.

"the person you want to be": Sam Anderson, "I Queue: Judging Your Friends by Their Netflix Lists," *Slate,* September 14, 2006. Available on the Web at http://www.slate.com/id/2149575/.

As AOL customers learned: Adam D'Angelo, "AOL Releases Search Logs from 500,000 Users," August 5, 2006, on the Web at http://www.ugcs.caltech.edu/~dangelo/aol-search-query-logs/. I've altered the data in this section.

165 *We don't even yet know if people will tag:* I heard this distinction between finding and refinding first from Thomas Vander Wal.

"an amplification system for memory": Joshua Schachter, telephone interview, January 2005.

166 *"to handle equivalence":* Peter Morville, *Ambient Findability* (Sebastopol, Calif.: O'Reilly, 2005), p. 139.

167 *Clustering tags:* I am on Technorati's board of advisers.

168 *"distribution and co-incidence of tags":* Email from Stewart Butterfield, March 14, 2006. The nose clusters are on the Web at http://flickr.com/photos/tags/nose/clusters/cat and http://flickr.com/photos/tags/nose/clusters/dog.

170 *We humans have a history:* Andy Clark, *Being There: Putting Brain, Body, and World Together Again* (Cambridge, Mass.: MIT Press, 1997).

9. MESSINESS AS A VIRTUE

173 *Gowns and street clothes:* Kendra Stanton Lee, "The Running of the Brides at Filene's Basement Is a Race for Wedding Dresses," *Associated Content,* March 2, 2006. Available on the Web at http://www.associatedcontent.com/article/22190/the_running_of_the_brides_at_filenes.html. See also Kim Knox Beckius, "The Running of the Brides," About.com, on the Web at http://gonewengland.about.com/od/bostonshopping/a/aarunningbrides.htm.

174 *In 1674, he boldly lumped:* Richard P. Smiraglia, "The Progress of Theory in Knowledge Organization," *Library Trends,* January 1, 2002. On Shakespeare variations, see David Kathman,"The Spelling and Pronunciation of Shakespeare's Name," on the Web at http://shakespeareauthorship.com/name1.html#2.

177 *"It is not enough to take this weapon":* Dwight D. Eisenhower, "Atoms for Peace," Address to the UN General Assembly, New York, December 8, 1953. Available on the Web at http://www.eisenhower.archives.gov/atoms.htm.

a story of power and fear: David Fischer, *History of the International Atomic Energy Agency: The First Forty Years* (New York: IAEA, 1997). Available on the Web at http://www-pub.iaea.org/MTCD/publications/PDF/Pub1032_web.pdf.

178 *led by West Point graduates:* Anton Chaitkin, "How the Government and Army Built America's Railroads," *Executive Intelligence Review,* July 17, 1998. Available on the Web at http://members.tripod.com/~american_almanac/railroad.htm. Alfred D. Chandler Jr., the author of one of the most highly regarded books on the history of management, thinks the influence of the military was more indirect: "Of the pioneers in the new managerial methods, only two—Whistler and McClellan—had military experience, and they were the least innovative of the lot." But, says Chandler, "Because the United States Military Academy provided the best formal training in civil engineering in this country until the 1860s, a number of West Point graduates came to build and manage railroads." He concludes, "There is little evidence that railroad managers copied military procedures." *The Visible Hand: The Managerial Revolution in American Business* (Cambridge: Belknap Press of Harvard University Press, 1977), p. 95. See also Edwin Lee Makamson, "The Rise of the Professional Manager in America," *The History of Management.* Available on the Web at http://www.mgmtguru.com/mgt301/301_Lecture1Page7.htm.

So he divided the company: Chandler, *The Visible Hand,* pp. 101–4.

180 *For Valdis Krebs:* Valdis Krebs, interview, March 20, 2006.

181 *"Ron Burt . . . studied Raytheon":* See Ronald S. Burt, "Social Origins of Good Ideas," unpublished paper, January 2003. Available on the Web at http://web.mit.edu/sorensen/www/SOGI.pdf.

182 "assuming a standard of virtue": Aristotle, *Politics,* Book 4, Part IX, translated by Benjamin Jowett, on the Web at http://classics.mit.edu/Aristotle/politics.4.four.html.

183 *Here there is no messiness:* On the problem of universals, see Gabora Aerts and Eleanor Rosch, "Steps Toward an Ecological Theory of Concepts," from a draft supplied by Dr. Rosch, p. 6. See also Eleanor Rosch, "Reclaiming Concepts," in *Reclaiming Cognition: The Primacy of Action, Intention and Emotion,* eds. R. Nunez and W. J. Freeman (Thorverton, U.K.: Imprint Academic, 1999), pp. 3–4. Published simultaneously as a special issue of the *Journal of Consciousness Studies,* vol. 6, no. 11–12 (1999), pp. 61–77.

"I think you unhorsed Aristotle": Interview with Eleanor Rosch, December 22, 2005.

184 *eleven basic color categories:* Philip E. Ross, "Draining the Language Out of Color," *Scientific American,* April 2004. Available on the Web at http://www.sciam.com/article.cfm?articleID=00055EE3-4530-1052-853083414B7F0000. John R. Taylor, *Linguistic Categorization,* second edition (Oxford: Oxford University Press, 1995). Taylor's book is an excellent introduction to Rosch's work. As Taylor notes, Berlin and Kay's work has been disputed.

185 *Wittgenstein's family resemblance:* Ludwig Wittgenstein, *Philosophical Investigations,* trans. G. E. Anscombe (Oxford: Blackwell, 1998), section 66.

186 *The basic-level names:* Brent Berlin, "Ethnobiological Classification," in *Cognition and Categorization,* ed. Eleanor Rosch and Barbara B. Lloyd (Hillsdale, N.J.: Lawrence Erlbaum, 1978), pp. 9–26.

(By the time children are four): George Lakoff, *Women, Fire and Dangerous Things: What Categories Reveal About the Mind* (Chicago: University of Chicago Press, 1987), p. 49.

187 *"Most if not all":* Rosch, *Cognition and Categorization,* p. 34.

In support, William Labov: Taylor, *Linguistic Categorization,* p. 40.

Research shows that for Americans: Geoffrey C. Bowker, "The Kindness of Strangers: Kinds and Politics in Classification Systems—Administrative History of Large-Scale Classification Systems," *Library Trends,* Fall 1998. He is citing *Linguistic Categorization* by John R. Taylor, pp. 44–57. See also Vivian Cook, "Words and Meanings," *Inside Language* (Oxford: Hodder Arnold, 1997). The prepublication word-processing file is available on the Web at http://homepage.ntlworld.com/vivian.c/Writings/Inside Language/ILvocab.htm.

188 *There was no set of attributes shared:* Eleanor Rosch and Carolyn B. Mervis, "Family Resemblances: Studies in the Internal Structure of Categories," *Cognitive Psychology,* vol. 7 (1975), pp. 573–605.

190 *"musty old book":* Tim Berners-Lee, *Weaving the Web* (San Francisco: HarperCollins, 1999), p. 1.

"Suppose all the information": Ibid., p. 4. Italics removed.

191 *In 1945, Vannevar Bush:* "As We May Think," *Atlantic Monthly,* vol. 176, no. 1, July 1945, pp. 101–18. Available on the Web at http://www.theatlantic.com/unbound/flashbks/computer/bushf.htm.

192 *Berners-Lee begins his* Scientific American *article:* Tim Berners-Lee, James Hendler, and Ora Lassila, "The Semantic Web: A New Form of Web Content That Is Meaningful to Computers Will Unleash a Revolution of New Possibilities," *Scientific American,* May 2001.

193 *A set of RDF triples:* Legal-RDF on the Web at http://www.hypergrove.com/legalrdf.org/index.html; http://www.hypergrove.com/legalrdf.org/inventory.html; and in "Legal XHTML: Event Classes," on the Web at http://www.hypergrove.com/OWL/Event/index.html #idActs; LRI Core on the Web at http://darius.lri.jur.uva.nl/wiki/ index.php/LRI_Core.

194 *MetaCarta:* In private conversation, June 2006. I have consulted to MetaCarta and am on its board of advisers.

The same is happening with the Semantic Web: For an excellent article on Berners-Lee's assessment of how well the adoption of the Semantic Web is going, see S. A. Mathieson, "Spread the Word, and Join It Up," *Guardian,* April 6, 2006. Available on the Web at http://technology.guardian.co.uk/weekly/story/0,,1747327,00.html. For a list of applications as of 2006, see Ivan Herman, "Tutorial on Semantic Web Technologies," on the Web at http://www.w3.org/People/Ivan/CorePresentations/RDFTutorial/Slides.html. Herman is with the W3C, the organizing body for Web technical standards.

It holds promise in health care: See W3C Semantic Web Health Care and Life Sciences Interest Group, on the Web at http://www.w3.org/2001/sw/hcls/; Eric K. Neumann, "Combining Drug Toxicity Knowledge," *Bio-IT World,* July–August 2006, on the Web at http://www.bio-itworld.com/issues/2006/july-aug/science-and-the-web.

195 *NeuroCommons.org:* See http://fm.schmoller.net/2006/05/semantic_web_in.html.

The Air Force Research Laboratory: Mark Gorniak, "Applying DAML to Foreign Clearance Guide," Presentation at Semantic Web for the Military User 2003 conference, May 7, 2003. Available on the Web at http:// www.daml.org/meetings/2003/05/SWMU/briefings/07_1355_AMC_FCG.ppt.

195 *development of the World Wide Web:* Thanks to Clay Shirky for suggesting this point, in conversation, September 2006.

"The Next Wave of the Web": Phil Windley, "The Next Wave of the Web," Technometria weblog, May 24, 2006, on the Web at http://www.windley.com/archives/2006/05/the_next_wave_o.

one of Japan's largest movie review sites: on the Web at http://microformats.org/wiki/hreview; calendar events on the Web at http://microformats.org/wiki/hcalendar.

196 *"The edges are fuzzy":* Interview with Joshua Schachter, December 22, 2005.

10. THE WORK OF KNOWLEDGE

199 *Harvard's Museum of Comparative Zoology:* Interviews, March 2, 2006.

201 *"collective delusion":* Clay Shirky, "Exiting Deanspace," Many2Many blog (to which I contribute), February 3, 2004, on the Web at http://many.corante.com/archives/2004/02/03/exiting_deanspace.php.

"an unlimited power to filter": Cass Sunstein, *Republic.com* (Princeton: Princeton University Press, 2002).

202 *"the one-way structure":* Yochai Benkler, *The Wealth of Networks: How Social Production Transforms Markets and Freedom* (New Haven: Yale University Press, 2006), p. 247.

204 *The claim that the Howard Dean campaign:* I was a Dean supporter and had a grandiose title with the campaign—senior Internet adviser—that overstates my contribution as a volunteer. For more about the campaign, see Joe Trippi's moving book *The Revolution Will Not Be Televised* (New York: Regan Books, 2004).

206 *Students of Marshall McLuhan:* For an application of this idea to the digitizing of information and communications, see Mark Federman, "Why Johnny and Janey Can't Read, and Why Mr. and Ms. Smith Can't Teach: The Challenge of Multiple Media Literacies in a Tumultuous Time," on the Web at http://individual.utoronto.ca/markfederman/WhyJohnnyandJaneyCantRead.pdf.

"Totemism is a subject": Harvey Einbinder, *The Myth of the Britannica* (New York: Grove Press, 1964), p. 41.

sizes of various encyclopedias: "Wikipedia: Size Comparisons," as of September 28, 2006, on the Web at http://en.wikipedia.org/w/index.php?title=Wikipedia:Size_comparisons&oldid=76919820.

Wikipedia's style guide: "Wikipedia: Article Size," on the Web at http://en.wikipedia.org/wiki/Help:Page_size.

207 *"One's encyclopedias grow less useful":* Einbinder, *Myth of Britannica,* pp. 270–71. Einbinder explains that originally "this was a letter in *Speaking of Holiday* (February 1961), a house organ of the Curtis Publishing Company."

207 *In the 1911 edition:* Tim Starling has scanned in the entire 1911 edition of the *Britannica*. The Goldsmith article is in volume 12, pp. 214–18. The figure of 6,000 words is based on an average of ten words per line. See http://en.wikisource.org/wiki/User:Tim_Starling.

209 *"I think that stuff is a hoot":* Interview with Jimmy Wales, January 2005.

Just a few hours later, over 2,400 bloggers: This is based on a search at Technorati.com using the key words *Bush* and *immigration*. Noted blog responses on the Web at http://www.purelyrandom.com/2006/05/16/amazing-bush-in-a-good-light, http://spaces.msn.com/blastfurnacecanada/Blog/cns!DB745086233C67DA!1068.entry, and http://www.alternet.org/blogs/peek/36305.

Siegel+Gale: On the Web at http://www.siegelgale.com/.

210 *a 2006 study by Edelman PR:* Press release from Edelman PR, January 23, 2006, on the Web at http://www.edelman.com/news/ShowOne.asp?ID=102. Disclosure: I consult for Edelman PR but had nothing to do with this study.

In the Istituto e Museo di Storia della Scienza: See http://brunelleschi.imss.fi.it/genscheda.asp?appl=SIM&xsl=catalogo&indice=54&lingua=ENG&chiave=407030.

Science . . . is not simpleminded: For a sympathetic exploration of science as both simplifying and complexifying, see Bruno Latour, *Pandora's Hope: Essays on the Reality of Science Studies* (Cambridge: Harvard University Press, 1999), pp. 70–71.

214 *"All the almanacs":* Interview with Bill McGeveran, February 2006.

215 Nature *magazine:* Nature Publishing Group History, on the Web at http://npg.nature.com/npg/servlet/Content?data=xml/02_history.xml&style=xml/02_history.xsl.

216 *"I wouldn't imply":* Interview with Philip Campbell, January 20, 2006.

a three-month experiment: "Nature Peer Review Trial and Debate," on the Web at http://www.nature.com/nature/peerreview/index.html.

Science *publication rates:* Alison McCook, "Is Peer Review Broken?" *Scientist,* vol. 20, no. 2 (February 2006), p. 26. Available on the Web at http://www.umkc.edu/research/listserv/6Feb06/The%20Scientist%20%20Is%20Peer%20Review%20Broken.htm.

At a site called arXiv: On the Web at http://arxiv.org. See Paul Ginsparg, "After Dinner Remarks," APS meeting at LANL, October 14, 1994. Available on the Web at http://people.ccmr.cornell.edu/~ginsparg/blurb/pg14Oct94.html. "Peer Review," Postnote, Parliamentary Office of Science and Technology, September 2002, no. 182, on the Web at http://www.parliament.uk/post/pn182.pdf. "The arXiv Endorsement System," on the Web at http://arxiv.org/help/endorsement. (Inevitably, some worthy papers are excluded; Yan Feng, a physicist at the European Southern

Observatory in Germany, noted that there weren't enough scientists in his field—complex photonics—publishing at arXiv to act as endorsers. Yan Feng, "The arXiv Endorsement System," April 25, 2005, on the Web at http://yanfeng.org/2005/04/the-arxiv-endorsement-system.) I found the Christopher Fuchs comment at the Quantum Pontiff blog, "Arxiv Links to Pontiff, Science at an End?" on the Web at http://dabacon.org/pontiff/?p=1189.

217 *Reddit.com:* Confirmed in an email from Aaron Swartz, March 2006.

Public Library of Science: From a conversation with Hemai Parthasarathy, October 20, 2006. A blog post about it is available on the Web at http://www.hyperorg.com/blogger/mtarchive/berkman_plos_open_access_scien.html.

219 *two most common words are* the *and* of: Rex Gooch, "Letter and Word Frequencies," *WordWays,* vol. 39, no. 2 (May 2006), pp. 98–99. The British word corpus can be found on the Web at http://www.natcorp.ox.ac.uk.

Tiger Woods responded: Gary Kamiya writes about Tiger Woods's embrace of his multiracial heritage and suggests it might influence Congress's upcoming decision to add a multiracial category to the 2000 census. Gary Kamiya, "Cablinasian Like Me," *Salon,* April 30, 1997. Available on the Web at http://www.salon.com/april97/tiger970430.html.

220 *U.S. Census categories:* See Colleen Monahan, "Using Census Data," University of Illinois at Chicago, on the Web at http://www.uic.edu/sph/dataskills/skillbytes/census/census3.htm; Robin Abrahams, "Censuring the Census," *Harvard Magazine,* March–April 2003, on the Web at http://www.harvard-magazine.com/on-line/030373.html; Annie E. Casey Foundation, "Using the New Racial Categories in the 2000 Census," on the Web at http://www.aecf.org/kidscount/categories/conclusions.htm.

Only 40 percent of those who declared: Kenneth Prewitt, a former U.S. Census director, speaking at Harvard University, February 2, 2004, as reported in Alvin Powell, "New Categories Cause Confusion," *Harvard University Gazette,* February 5, 2004, on the Web at http://www.news.harvard.edu/gazette/2004/02.05/15-census.html.

"has no scientific justification in human biology": "American Anthropological Association Response to OMB Directive 15: Race and Ethnic Standards for Federal Statistics and Administrative Reporting," September 1997, on the Web at http://www.aaanet.org/gvt/ombsumm.htm.

members of a race differ among themselves genetically: M. Nei and A. Roychoudhury, "Gene Differences Between Caucasian, Negro, and Japanese Populations," *Science* 177 (1972), pp. 434–35.

229 *the platypus could have:* Harriet Ritvo tells well the story of the reluctance of nineteenth-century taxonomists to accept the platypus as a real animal. See the first chapter of her excellent *The Platypus and the Mermaid and Other Figments of the Classifying Imagination* (Cambridge: Harvard University Press, 1997).

NOTES

1. I am on the advisory boards of some companies I mention in this book because I want to support companies I believe are improving the Internet ecology. I recognize the conflict and have tried to address it by examining my motives and by noting the relationship in the endnotes. I also have bonds of friendship with some of the people who show up in this book. It is an intertwingled world, and I have tried not to let those relationships sway my judgment but, of course, no human can entirely escape such influences.

2. You may have noticed that I use the feminine pronoun rather than the masculine or the "she or he" work-around. That's because for the first half of my life, I assumed the masculine. For the second, I've been assuming the feminine.

3. The book has a Web site—http://www.EverythingIsMiscellaneous.com—that has more information and an open discussion of the book's topics. You'll also find the bibliography there, in which I try to make up for consigning to the footnotes towering thinkers such as Michel Foucault.

THANK-YOUS

Many people have been of continuing help throughout this book's long period of gestation and development.

The Harvard Berkman Center for Internet & Society has twice renewed my fellowship there, making the resources of that university available to me. Even more important, it put me in the warmest, most caring academic environment I've encountered, where a remarkable community of minds have steered my thinking. Thank you to my fellow fellows.

Other individuals have been unstinting in their help. Clay Shirky is a model of generous thinking. His insights and clarity—and his friendship—have helped me from the very beginning of this project. David Miller, my agent and friend, over years of gluten-based meals has been crucial in helping me sift through the possible ways of approaching this topic, and has in conversation challenged and extended my thinking. Robin Dennis has exceeded my every hope of what an editor could be, criticizing, refining, and improving the words and ideas in this book. I am also indebted to Professor Joseph Fell, who introduced me to philosophy when I was a freshman at Bucknell University. Thirty-eight years later, Professor Fell took time away from his own crucial work on the problem of universals to give an early draft of this work a thorough reading. In addition, some people I deeply respect read early drafts and commented generously and helpfully; a special thanks goes to Kevin Marks. Thanks to *Wired, Harvard Business Review*, and *Release 1.0*, which ran articles of mine on some of the themes discussed in this book. I have also blogged various ideas and requests for help, and bloggers have responded in every case in helpful and surprising ways. Thank you for the social thinking.

I owe my family a warm and deeply felt thank-you, especially our children who may have teased me roundly whenever they heard the word "taxonomy" forming in my mouth but then jumped right into the conversation. My wife, Ann Geller, was always ready to put aside her own studies to listen to a draft or a drafty idea. There's no one who knows Ann who doesn't love her; I am the prototype of that truth.

Finally, Danny Danziger, and John Gillingham, coauthors of *1215: The Year of the Magna Carta,* end their introduction by saying that any errors are the other author's fault. I would like to say the same thing, but since I am the sole author of this book, I have no choice but to admit that any errors are the fault of Danny Danziger and John Gillingham.

INDEX

ABOUT THE AUTHOR

DAVID WEINBERGER is the coauthor of the international bestseller *The Cluetrain Manifesto* and the author of *Small Pieces Loosely Joined.* A fellow at Harvard Law School's Berkman Center for the Internet & Society, Weinberger has written for such publications as *Wired, USA Today, Smithsonian,* and the *Harvard Business Review* and is a frequent commentator on NPR's *All Things Considered.* As a marketing consultant, he has worked with Fortune 500s, leading media companies, and many innovative start-ups, and he served as senior Internet adviser to the Howard Dean presidential campaign. Weinberger holds a doctorate in philosophy. He lives in Boston.